Technology and Society

Technology and Society

Social Networks, Power, and Inequality

Anabel Quan-Haase

OXFORD
UNIVERSITY PRESS

OXFORD
UNIVERSITY PRESS

Oxford University Press is a department of the University of Oxford.
It furthers the University's objective of excellence in research, scholarship,
and education by publishing worldwide. Oxford is a registered trade mark of
Oxford University Press in the UK and in certain other countries.

Published in Canada by
Oxford University Press
8 Sampson Mews, Suite 204,
Don Mills, Ontario M3C 0H5 Canada

www.oupcanada.com

Library and Archives Canada Cataloguing in Publication
Quan-Haase, Anabel Technology and society:
social networks, power, and inequality / Anabel Quan-Haase.

(Themes in Canadian sociology) Includes bibliographical references and index.
ISBN 978-0-19-543783-6

1. Technology—Social aspects. I. Title. II. Series: Themes in Canadian sociology

T14.5.Q82 2012   303.48'3   C2012-902502-X

Cover image: Volodymyr Grinko/iStockphoto

Oxford University Press is committed to our environment. This book is printed on
Forest Stewardship Council® certified paper and comes from responsible sources.

Printed and bound in Canada

1 2 3 4 — 16 15 14 13

Contents

Preface

A key motivation for writing *Technology and Society* was a mixed sense of euphoria and concern. On the one hand, as I continued to adopt various technologies and to test new applications on the Web, I felt excited to be witnessing a time of unprecedented technological transformation—the era of *digital tools*. This is an era where strings of bits and bytes have opened the door to endless combinations that yield a large proliferation of new tools, such as cellphones, e-book readers, and social media. I had come to rely on some of these ubiquitous technologies as if they had always been there, as they allowed for flexibility, mobility, and connectivity. On the other hand, I wondered what would happen if these technologies failed us. My real concerns were not about some doomsday predictions, where the world would come to a sudden halt, but, rather, about my sociological curiosity to understand what these technologies meant for society as a whole.

The purpose of this book, then, is to slow down and step back for a moment to make technology the object of sociological inquiry and to uncover the intricacies of our socio-technical life. By taking a socio-technical perspective, the book makes readers aware of the pervasiveness of technology in our everyday lives and encourages an understanding of how technology interacts with and is embodied in society. Technology is both the driving force behind societal change as well as the output of our technological imagination. It is this dichotomy that we want to present. The focus of the book is on the high dependence on all things technological, combined with the problems, social issues, and socio-political realm in which these technologies are embedded. As Marshall McLuhan so eloquently pointed out, technology is not neutral. It affects our society and we need to be able to discuss and scrutinize these effects. This book provides the necessary background to start such an analysis by defining the parameters around technology.

The book as a whole has three goals. The first goal is to examine how technology and society intersect. The book investigates how technological change is interlinked with inequality, power, and social networks. It is about connecting issues of relevance at both local and global, micro and macro levels. In exploring these connections, this book raises and attempts to answer a number of central questions: How is technology leading toward social change? What is the nature of this social change? Who is being affected by the technologies? Are various social groups being affected differently? How are people using the technologies in different regions? Does the global digital divide continue to exist?

The book's second goal is to draw on readers' own experiences as a means to make sense of the link between technology and society. What is unique about this book is its focus on experiential learning, which emphasizes meaning from direct experience. Aristotle was among the first to recognize the value of experience in gaining knowledge: "For the things we have to learn before we can do, we learn by doing" (Bynum & Porter, 2006). The book serves as a bridge, then, between abstract, theoretical thinking and real-life events and experiences. Throughout the book, on-the-ground examples are utilized to demonstrate the relevance of concepts and ideas. These real-life examples also demonstrate the relevance of studying technology in its specific social context, be it historical, geographic, regional, linguistic, age-based, or ethnic. Technology is about context. To understand better the contextual nature of technology, the book strongly encourages students to bring their personal experiences and voices to bear on the material. This helps to put the concepts and theories in perspective and give them meaning, immediacy, and relevance.

The final goal of the book is to encourage curiosity and to find out more about technologies. The book should give readers an urge to do something about the material: share it, post it, comment on it, and make a difference in the world. Technology has become such an intrinsic part of our everyday lives in the West that we need to care about these social problems and the questions that open from the critical analysis of text, theory, and experience.

What is difficult about any study of technology is to step back and look at technology as neutral, critical observers. But doing so allows us to examine technology in many different ways: technology as a technical device, as a social force, as an agent for warfare, as a tool for health care, as a device for fun, or as a force behind job losses and deskilling. A great example of stepping back and just looking at technology is Woody Allen's movie *Sleeper*, which shows how ridiculous our reliance on technology can become. In one scene, two tailor robots fit Miles Monroe (played by Allen) with a suit and the best they can come up with is an oversize, ugly grey coat. Annoyed, Monroe responds, "This is terrible." The robot is willing to accept the customer's complaint: "Oh well, oh well, we will take it in." In this case, even the robot can see how silly the entire process is. The movie represents both a compelling picture of the role of technology in the future as well as a satire of our visions of, beliefs in, and reliance on technology.

Can we actually step back as neutral observers? It is hard to imagine a world without technology. Indeed, technology has been a part of human existence since the Stone Age when humans used stones, bones, and sticks as tools for survival. But studying technology can help us become more aware of its role in our personal lives, in the lives of other social groups and their struggles, and in our society as a whole. Moreover, systematic analysis design, implementation, and use allows us to develop theories of the intersection of technology and society. This type of analysis provides us with necessary background knowledge and tools to embark on social, socio-political, and cultural studies of our own socio-technical existence. As a result, this book provides a solid understanding of technology's role in society and gives students the tools they need to embark on a critical and in-depth inquiry of our technological society.

Overview of the Book

Technology and Society: Social Networks, Power, and Inequality is aimed at undergraduate courses on technology in a range of disciplines in the social sciences (in particular sociology and anthropology), arts and humanities, communication studies, and information science. This book is also appropriate for management or business classes because of its focus on technology design, innovation, and labour in Web 2.0 contexts. The book does not require any previous knowledge of technology or statistics. Theories and concepts are explained in great depth, and the glossary provides definitions of new and specific terminology. *Technology and Society* relies on current interdisciplinary work from sociology, the history of technology, science and technology studies (STS), communications, and related fields.

The chapters are organized to help students understand and learn the material. Each chapter starts with a set of learning goals, continues with a general overview or introduction to guide the readers through the material, and ends with a conclusion, a set of study questions, and further readings. Through its comprehensive list of further readings and additional Web content, the book encourages readers to seek out further resources, to obtain additional current information, to deepen their knowledge of topics, and to explore new topics of their own interest. Next, we present a short overview of each chapter to give readers an idea of some of the topics covered.

Chapter 1: The Technological Society

Chapter 1 investigates the contentious question of how to best define *technology*. The chapter outlines and critically discusses several approaches. Considering the depth and pervasiveness of technology in our society, this introductory chapter stresses the relevance of studying the intersection of technology and society—what is often referred to as the *socio-technical perspective*. The chapter includes a discussion of how technologies lead to large-scale, widespread social change, such as social and economic inequality. The key argument is that social change occurs not as a result of technology alone but, rather, as a blending of micro-, meso-, and macro-level processes. In addition to established approaches to understanding technology, the chapter also covers two of the main contemporary perspectives surrounding the use of technology in modern society. The first is simulation, which is geared toward the development of tools that can resemble or outperform human faculties. The second perspective is augmentation, which attempts to integrate machines and humans into new hybrid actors with added capabilities. Both of these perspectives highlight the many points of intersection of the human and the technological, raising important questions about the ethical, moral, and societal implications of endeavours such as augmentation and simulation.

Chapter 2: Technology in Society: A Historical Overview

To comprehend fully how technology and society intersect in our modern society, we need to first take a look at the history of technology. The aim of Chapter 2 is to provide a broad overview of this history by tracing the roots of technological development, discussing key periods of technological innovation, and outlining our present-day, high-tech society. Technology is the strongest force of change in society and, as such, its development, transformation, and diffusion directly shape many aspects of society, such as work, community, and social relationships. We examine these technological transformations over time and demonstrate the impacts they have had on past societies. This chapter further attempts to link technologies, inventors, and historical moments to provide an in-depth examination of the socio-political context in which technologies emerge. Chapter 2 shows how technology is ingrained in society, affecting all aspects of our lives, and it outlines the merits of taking a sociological perspective when studying the history of technology.

Chapter 3: Theoretical Perspectives on Technology

Chapter 3 covers a plurality of theoretical perspectives, which seek to shed light on the nature of the relationship between technology and society as well as on those elements of society most affected by technology. First, the chapter contrasts the utopian and dystopian approaches, which each highlight a different side of how technology transforms society. Second, Chapter 3 also reviews the key premises underlying the theories of technological and social determinism and discusses their strengths and weaknesses. Third, the chapter introduces the field of STS as an alternative framework, which stresses that artifacts are socially constructed, mirroring the society that produces them. Then the chapter reviews the three most prominent approaches of STS: actor network theory (ANT), social construction of technology (SCOT), and social informatics (SI).

Chapter 4: Techno-Social Designing

The topic of Chapter 4 is the impact of social factors on the design of technology. Users of technology are often oblivious of the complexity underlying technological design because not much knowledge is available on how this process unfolds. The aim of this chapter, then, is to uncover these often hidden creative processes by examining developers' visions and the challenges experienced in research and development (R&D), including the pressures that exist in R&D teams and the ways that innovation occurs in these teams. As well, Chapter 4 introduces the term *technopole* to describe specialized cities dedicated solely to technological innovation. Silicon Valley, for example, is presented as an example of a technopole that combines a highly educated workforce with military and economic interests. The chapter also examines in more detail the inner workings and the outer pressures of software development, which is one type of R&D that has come to occupy a central role in the world economy.

Chapter 5: The Adoption and Diffusion of Technological Innovations

Chapter 5 investigates how technology diffuses in society. The chapter provides an overview of the key concepts, theories, and research findings in the diffusion of innovations literature and discusses them in relation to the diffusion of specific technologies, such as water boiling, the QWERTY keyboard, and the iPhone. The aim of the chapter is to examine key adopter groups, how they differ, and their salient characteristics. Everett Rogers' classic model of the diffusion of innovations, his model of the innovation-decision process, and his categorization of adopter groups are elucidated in great detail to provide students with the necessary foundation. As part of this overview, the roles of change agents and early adopters are discussed as being among those who play an active role in promoting and diffusing new tools.

Chapter 6: The Labour of Technology

The complex interrelation between technology and labour is the topic of this chapter. The focus here is on the changes in the nature, context, and structure of work that result from new technologies and the associated power imbalances. Historically, these changes occurred as early as the Industrial Revolution when machines replaced skilled workers, creating social upheaval, dissatisfaction, and unrest. We use the historical context to better grasp current trends in how Web 2.0 technologies facilitate new forms of production that are based on principles of collaboration, sharing, and open source. We discuss key concepts of the Web 2.0 mode of work, including *prosumer, produsage,* and *perpetual beta.* Wikipedia and Facebook are used as examples to illustrate current trends in how users become producers of content. Most users would not consider this work but, rather, art, pleasure, fun, and leisure time. However, these new forms of work do have consequences for the new economy and labour relations. The main point of this chapter is that technology is not neutral but, rather, becomes an active force that changes the nature of work itself, working conditions, and the structure of society as a whole.

Chapter 7: Technology and Inequality

In Chapter 7 issues specifically linked to digital inequality are covered to provide an overview of the social, economic, and cultural consequences that result from a lack of access to the Internet. The chapter covers the historical developments of the digital divide concept, examines the complexity of its measurement, and considers its relevance to policy in Canada and the United States. A key argument is that inequalities in the use of digital technologies reflect not only problems associated with access to networked computers but also differences in skill level among users. Another central term covered in this chapter is the *global digital divide*, which describes the gap in access to the Internet that exists between developing and developed nations. Many developing nations continue to falter in their efforts to become digital, having to overcome numerous barriers of access, skill, and infrastructure. China is discussed as a prime example of a newly industrializing nation that has struggled to join the information society and in the process has developed an ambivalent relationship with the Internet.

Chapter 8: Community in the Network Society

Chapter 8 examines how the notion of community has changed as a result of the introduction of new technologies in society. The chapter starts with a brief overview of *Gesellschaft* and *Gemeinschaft* as two of the most central concepts in the study of the structure of society. What follows is a critical examination of the debate about how industrialization, urbanization, and globalization have affected community. The community-lost, community-saved, and community-liberated perspectives are reviewed to contrast competing theories on the nature of these changes. The chapter also considers the recent concerns expressed about the impact of the Internet on the patterning of social relationships and presents various competing perspectives. Chapter 8 concludes with a discussion of how information and communication technologies have affected the public sphere. As part of this discussion, we take an in-depth look at the events that unfolded in Egypt in February of 2011 and analyze the role of social media in the protests.

Chapter 9: Technology-Mediated Social Relationships

In Chapter 9, we briefly outline the early beginnings of mediated communication and address how they impact society. Then, we review recent trends in how people form and maintain personal connections via the Internet. While most discourse focuses on the benefits of mediated communication, scholars have also warned about the potential negative effects on people's social life. Can personal relations maintained online provide as much social support as those maintained in person? The chapter then focuses on how social media have redefined our notion of friendship and the implications of these changes for community and social networking. What follows is an analysis of romance 2.0, investigating how people form and terminate romantic relations using social media. The chapter ends with an exploration of the concept of virtual mourning and how people renegotiate, online, the meaning of death. To explore this new concept in depth, we examine the unprecedented online response to Michael Jackson's death and the supportive online community that formed around Canadian Eva Markvoort during her struggles with cystic fibrosis.

Chapter 10: The Surveillance Society

The topic of Chapter 10 is surveillance and how it has become a central concern of our digital age. The goal of this chapter is to define the multi-faceted term by contrasting different perspectives available in the literature. The chapter then provides an overview of the concept and architecture of the Panopticon and its means of exerting control and imposing disciplinary action. The chapter discusses the new modes of surveillance made possible by recent technological developments, which show how technologies have not only changed the practices of surveillance but also the very nature of surveillance, to some extent largely reducing individuals' privacy rights. We end the chapter with a discussion of innovative methods of counter-surveillance that aim at increasing awareness of the pervasiveness of surveillance in our society and provide means for personal resistance.

Chapter 11: Ethical Dimensions of Technology

The goal of the final chapter is to summarize the three key themes that run through the book. The first theme stresses the need to take a socio-technical perspective that allows for an in-depth examination of the social context of technology design, use, and implementation. The second theme demonstrates how innovation is associated with economics and as a result has consequences for our understanding of inequality and power relations. The last theme shows the many ways in which technological developments lead toward social change, impacting community, social networks, and social relations. The final chapter also embarks on a critical examination of the ethical and moral dimensions of humans' engagement with technology. The following themes are explored: the neutrality of technology, technology as human destiny, and technology as progress. Through this discussion, the chapter emphasizes the unexpected consequences of technology that society needs to come to terms with.

Acknowledgements

I am very grateful to Lorne Tepperman and Susan McDaniel for giving me the opportunity to contribute to the Themes in Canadian Sociology series; this has been a great honour and privilege. I would also like to thank Jodi Lewchuk and Mark Thompson at Oxford University Press for all of their advice and encouragement during the writing and editing of the book. The manuscript has benefited enormously from the constructive and helpful feedback received from Barry Wellman at the University of Toronto and Laurie Forbes at Lakehead University. I thank them for their time and commitment to high standards in scholarship. This project would not have come to fruition without the love of my family—the Mortons and the Quans. They bring many of the examples in the book to life with their devotion to and skepticism of all things technology. I would also like to express my indebtedness to my research assistants—Becky Blue, Gary Nicholas Collins, Michael Haight, Kim Martin, and Jonathas Mello—who helped with literature reviews, drafts, editing, graphics, and humour. Chapter 1 greatly benefited from discussions with my colleague Victoria L. Rubin, whose sharp insights helped reorganize the chapter. Chapter 6 benefited from discussions with Brian Brown, while Chapter 10 received input from surveillance expert John Reed.

Abbreviations

AI	artificial intelligence
ANT	actor network theory
CANARIE	The Canadian Advanced Network and Research for Industry and Education
CAP	Community Access Program
CBC	Canadian Broadcasting Corporation
CCTV	closed circuit television
CF	cystic fibrosis
CGE	Compagnie Générale d' Electricité
CMC	computer-mediated communication
CNNIC	China Internet Network Information Center
DD	digital divide
DIY	do-it-yourself
EDF	Électricité de France
FCC	Federal Communications Commission
GDP	gross domestic product
GERD	gross domestic expenditure on research and development
GSS	General Social Survey
ICTs	information and communication technologies
ICT4D	information and communication technologies for development
IM	instant message
ISP	Internet service provider
IT	information technology
ITU	International Telecommunications Union
KME	Knowledge Management Enterprises
LAN	local area network
LDC	least developed country

NPOV	neutral point of view
NRA	National Rifle Association of America
NTIA	National Telecommunications and Information Administration
OECD	Organisation for Economic Co-operation and Development
OLPC	One Laptop per Child
PC	personal computer
R&D	research and development
ROI	return on investment
SCOT	social construction of technology
SI	social informatics
SNS	social network sites
STS	science and technology studies
UNDP	United Nations Development Program
WELL	Whole Earth 'Lectronic Link
WSIS	World Summit on the Information Society

1

The Technological Society

Learning Objectives

⊛ To understand the challenges in defining and studying technology.
⊛ To discuss the relevance of studying the intersection of technology and society.
⊛ To learn about contemporary perspectives of technology.

Introduction

This introductory chapter highlights the relevance of investigating the intersection of technology and society. It is easy to dismiss technology as a mere object, without giving much consideration to how it is woven into our everyday lives. Technology provides a means for us, its users, to get things done. Without giving it much thought, we leave our homes every morning with our cellphones (often more than one), laptops, MP3 players (such as iPods), headphones, watches, and other gadgets. Only when our technology fails us do we suddenly realize the depth of our dependence on that technology. There is frustration and sometimes even a sense of panic when we forget our cellphone at home (or worse, when we lose it!). Our cellphone's digital address book has become an extension of our memory, storing hundreds of names, phone numbers, and email addresses. This raises questions about the use of technology as a means to enhance, complement, or even substitute human faculties. As long as technology works smoothly, it is simply part of our daily life, of how society ought to function. But when it fails, disaster can follow. This reality became starkly apparent in 2011, when the Tōhoku tsunami crashed into Japan causing massive structural damage and revealing the country's high dependency on, and vulnerability to, nuclear energy (and its potential collapse). Until then, nuclear energy had been a given, a part of how Japanese society functioned. Now questions are being raised about whether nuclear energy is an acceptable means to an end.

The aim of this initial chapter is to examine how the social and the technological intersect, which is referred to as the **socio-technical perspective**. To introduce readers to the topic, the chapter critically discusses and compares various definitions of *technology* and ends with a working definition that guides the remainder of the book. In addition, the chapter introduces and contrasts two perspectives of technology that define our current times. The first is simulation, which is geared toward the development of tools that can resemble or outperform human faculties. The second perspective is augmentation, which attempts to integrate machines and humans into new hybrid actors with added capabilities. Both of these perspectives highlight the many points of intersection of the human and the technological, raising important questions about the ethical, moral, and societal implications of endeavours such as augmentation and simulation.

Challenges in Defining and Studying Technology

In the 2009 movie *Star Trek*, Scotty suggests that "transwarp beaming" is "like trying to hit a bullet with a smaller bullet, whilst wearing a blindfold, riding a horse" (Abrams, 2009).[1] Defining *technology* presents a similar challenge as scholars have proposed many different definitions of the concept. Additionally, in Box 1.1 we will examine in more detail the challenges inherent in studying technology.

What Is Technology?

In this section, we will review five different definitions, from the narrowest to the broadest, which view technology as (1) material substance, (2) knowledge, (3) practice, (4) technique, and (5) society.

1. Technology as Material Substance

Until the nineteenth century, the study of technology was primarily the focus of the technical fields or applied sciences; little work on technology was done in, for example, the social sciences or humanities. In technical fields technology was primarily defined as consisting of technical components; little attention was paid to the interplay of technology and society. Within this very materialistic approach, technology is viewed as "a radical other to humanity" (Feist, Beauvais, & Shukla, 2010, p. 8), an entity that exists outside of the social realm. This approach sees technology, then, as a passive object, a tool created by humans to be used under our control (Feist et al., 2010).

By examining technology as only **material substance**, the complex interplay of technology and society is disregarded. This tunnel vision limits our understanding of technology and does not allow for an examination of social change resulting from technology. This view has been largely discredited and a number of alternative definitions have been proposed.

2. Technology as Knowledge

In the same way that technology is closely interlinked with science, it is also closely connected to knowledge. At the most basic level, "[t]echnology is based upon, utilizes, and generates a complex body of knowledge, part of which may reasonably be called specifically technological knowledge" (McGinn, 1978, p. 186). In contrast to scientific knowledge, technological knowledge stems from human activity, often in relation to an **artifact** (Hershbach, 1995). *Artifacts*[2] are defined as all objects that have been modified, modelled, or produced according to a set of humanly imposed attributes. Examples of artifacts are tools, weapons, ornaments, utensils, and buildings. Technological knowledge is therefore focused on the ability to create, utilize, or transform objects with the aim of facilitating certain activities or achieving specific goals (McGinn, 1978). This distinguishes technological knowledge from scientific knowledge, which can be applied to the design and building of artifacts but is not directly linked with practical applications (see the discussion in Chapter 2). Scientific knowledge, in contrast, is abstract and consists of our understanding of the natural world. For instance, calculus, as a branch of mathematics, provides an understanding of limits, functions, derivatives, and integrals, and at the same time we can observe its value in real-world applications. For instance, calculus provided to a large extent the foundation for a practical **invention**, namely the introduction of the steam engine (Restivo, 2005).

The metaphor of technology as knowledge can, however, be limiting in two ways. First, it does not consider that the knowledge required to create, utilize, and transform objects is a different entity than the object itself, even if the object was invented based on this knowledge. Second, the metaphor disregards the impact that technology has on society by limiting technology to expertise, skill, and know-how. Layton's model of technology and knowledge is useful to examine next because it overcomes some of these limitations.

While also using the metaphor of technology as knowledge, in his model Layton distinguishes between the technology itself and the knowledge available about the technology. At the centre of **Layton's model of technology** is the process of technological development in which "technological ideas must be translated into designs" which then "must be implemented by techniques and tools to produce things" (Layton, 1974, p. 38). From this viewpoint, a model of technology emerges where technology is not a single entity but, rather, embodies three different elements:

1. *Ideas*: Ideas are at one end of the spectrum representing the thought processes that precede the tool, which is often referred to as the *technological imagination*. Ideas about what objects are useful motivate design and development.

2. *Design*: Design is in the middle of the spectrum, mediating between the abstract idea and the object. Design is the step that is required to go from idea to tool development and is where craftsmanship is needed.

3. *Techniques*: Technique describes the actual artifact, and lies at the other end of the spectrum. The artifact is made up of material substance and allows humans to complete tasks.

Layton's model of technology has two advantages. First, the definition of *technology* goes beyond a pure consideration of artifact. The model explicitly distinguishes between the artifact and the knowledge from which it arises. Second, the model also incorporates design as a middle stage that lies between the idea of a tool and the artifact itself (discussed in Chapter 4). This creates a link between how we envision tools and their actual realization, which are often two different things. One disadvantage of the model, however, is that there is no explanation of how technology and technological knowledge are linked to society. In this view, technology remains conceptualized as an artifact, only drawing on knowledge and design and in isolation from culture, economics, power, etc. That is, as in other theories of technology as knowledge, the socio-technical is largely disregarded.

3. Technology as Practice

A third set of definitions have broadened the meaning of technology. The Canadian scholar Ursula Franklin (1992) has chosen to define *technology* in the context of the real world. For her, technology is not limited to the apparatus, to the material substance, or to the artifact. Nor is it just "the sum of the artifacts" but instead "a system" involving organization, procedures, symbols, new worlds, and most importantly, a mindset (p. 2). Hence, technology needs to be understood as practice, as embedded in the everyday activities of humans.

Franklin's view is rather pessimistic in that the "real world of technology seems to involve an inherent trust in machines and devices ('production is under control') and a basic apprehension of people" (p. 25). Franklin criticizes society's simplistic view: while people create problems and are unpredictable, machines provide solutions to problems and are always controllable (1992). Hence, technologies are always idealized because any problems that arise from technology can be easily blamed on people, on those who designed, produced, or consumed technology. For Franklin (1992), a technology that is often perceived as being able to "liberate its users," often ends up enslaving them by creating a dependence on that technology. For instance, the iPod allows users to listen to music on the go and, consequently, creates a barrier between users and their external environment. Franklin argues that these tools have contributed to a world of "technologically induced human isolation" (p. 46).

Instead of conceptualizing technology as an external force that functions outside of the everyday realm, viewing technology as practice places technology within people's everyday lives. This is the strength of this approach; it emphasizes how technology becomes normalized within society over time and thereby points out how technology is a structuring force in how we live, play, and work. Simply put, we do not question our routines but, rather, see them as part of our everyday lives. When we put on our running shoes and tie our laces, this is a *normalized* behaviour. We do not realize that shoes are an invention (as the Bata Shoe Museum in Toronto documents) and that shoelaces are a complex technology that developed over centuries (Tenner, 2003).

The limitation of Franklin's view is that technology is analyzed primarily as a negative force. From this viewpoint, it is technology that determines how social change occurs in society, and it neglects to examine human agency (see Chapter 3 for a discussion of technological determinism). Such an interpretation limits how we view technology because it does not allow for people to make decisions about which technologies they want to use, how they want to use them, and how these technologies affect their lives.

4. Technology as Technique

A different approach to defining *technology* comes from scholars who look at the problem from an action-oriented point of view. These thinkers aim to identify the essence of technology by distinguishing between *technology* and *technique*, which is a translation of the German word *Technik* (Hanks, 2010). **Technique** then describes an abstract concept and not an object. Martin Heidegger (2010) describes *technique* as a human activity and further divides it into **goal** (*Zweck* in German) and **mechanism** (*Mittel* in German). The goal is that which humans want to achieve with the technique. It can be an activity or a more abstract state of existence. The mechanism provides the means to realize the goal—this includes materials, procedures, know-how, and societal norms. Heidegger argues that the activity of technique and its products have an impact on every aspect of life.

Jürgen Habermas also describes technology in terms of the realization of goals. He sees technology as **strategic action** because it provides the means for the realization of human endeavours (Simpson, 1995). Habermas (1984) identifies a strategic action as a social action that is "oriented towards success" and that has considered "the aspect of following rules of rational choice and assess[ing] the efficacy of influencing the decisions of a rational opponent" (p. 285).

The emphasis on technology as goal was also present in Jacques Ellul's study of technology, where he defined *technique* as a standardized means for attaining a predetermined goal in society (Merton, 1964). Ellul also described technique as a mechanism characterized by efficiency: "the totality of methods rationally arrived at, and having absolute efficiency (for a given stage of development) in every field of human activity" (1964, p. xxv).

Ellul (1964) was interested in examining the lasting effect of efficient techniques and in this context he viewed technique as the "defining force of a new social order in which efficiency is no longer an option but a necessity imposed on all human activity" (p. 17). His critique of technique focuses on the intricate link between technology and its social, psychological, moral, and economic consequences, which are not necessarily in agreement with human ethics. Hence, according to Ellul, technology imposes the principle of efficient ordering on society at the expense of other considerations.

In these conceptualizations of technology as technique, technology is examined in relation to human activity. Technology is no longer "just a tool"; it becomes a mechanism for achieving human needs and wants through efficient systems. While these thinkers take our understanding of technology a step further and acknowledge the lasting impact of technology on society, they do not elaborate on the mechanisms by which social change occurs. Again, technology continues to be an agent, a force external to society.

5. Technology as Society

When we define *technology* as material substance, knowledge, practice, or technique, we limit our ability to analyze the technological in society. Definitions of *technology* need to be broader to allow for a comprehensive examination of its close interplay with society. Narrow definitions limit our ability to see the realms in which technology has left a mark. Simpson (1995) provides a comprehensive definition, stating that technology encompasses knowledge, mechanisms, skills, and apparatus that are geared toward controlling and transforming society. In this perspective, technology goes beyond merely being machinery or practice—it becomes an agent of change that can control and alter humanity.

Jean Baudrillard goes even further by arguing that technology does more than effect change in society: "[i]t doesn't push things forward or transform the world, it becomes the world" (Baudrillard & Gane, 1993, p. 44). Baudrillard is not the only one to argue that technology and society become one and the same entity. For Herbert Marcuse, technology is "a social process, in which technics proper (that is the technical apparatus of industry, transportation [and] communication) is but a partial factor" (1982, p. 138). As part of his critique of consumerism, Marcuse believes that the introduction of new technologies brought about new standards and cultural and social change, which were not purely the effect of technology but "rather themselves determining factors in the development of machinery and mass production" (1982, p. 138). This critique is perhaps one of the most relevant analyses of the technology–society link because it highlights the complex interdependence between cultural and social change and technological developments.

As with Baudrillard, Marcuse points to diminishing critical thought in society as the clamour for material goods increases. The radical view that technology is equal to society points out that our technological society cannot exist outside the framework of technology. Technologies are not only ubiquitous in our daily lives (e.g., transport, communication, food provision, and consumption), but even our way of seeing the world is mediated by technology. There is no escaping it. However, this view is also problematic because it does not allow us to study how technology intersects with society. In order for us to be able to study technology, we need to define it as something separate from society itself. If our definition of technology argues that it has become so intertwined with society that the two are in fact a single unit, this precludes any kind of serious investigation of how they influence each other.

Making Sense of These Definitions

It is important to understand the range of existing definitions of technology because this will provide the basis of our discussions of how technology intersects with society in the remainder of the book. Each of the five different definitions of technology reviewed has its merits, helping us understand one aspect of technology. Nevertheless, it is equally important to realize that no single definition can fully capture the meaning of such a complex concept. Hall has argued that we have to recognize the rich technological bases of modern cultural production, which enable us to endlessly simulate, reproduce, reiterate, and recapitulate. But there is all the difference in the world between the assertion that there is no one, final, absolute meaning of technology and the assertion that meaning does not exist (Grossberg, 1996, p. 137).

The definitions of technology as material substance, knowledge, practice, and technique are all rather limiting and do not show sufficiently how technology links to society. On the other hand, the definition of technology as society is broad in scope and directly attempts to understand how society and technology come together. The problem, though, with broad conceptualizations such as those of Baudrillard is that they fuse society and technology into one and the same thing: "[i]t doesn't push things forward or transform the world, it becomes the world" (Baudrillard & Gane, 1993, p. 44). This makes it impossible to study how the two are linked and also neglects to study the technology–society interdependence.

The definition of technology that we will adapt in this book is simple and draws on elements from the five definitions discussed above:

> Technology is an assemblage of material objects, embodying and reflecting societal elements, such as knowledge, norms, and attitudes that have been shaped and structured to serve social, political, cultural, and existential purposes.

Despite its simplicity, the definition has three advantages: (1) it separates technology from society sufficiently enough to allow for an investigation of the technology–society interdependence; (2) it does not include technology as an element of society but, rather, sees technology as serving social purposes; and (3) it views technology as embodying and reflecting knowledge, social norms, **social structures**, political interests, and so forth. This definition will guide our inquiry in the remainder of the book.

We conclude that *technology* has multiple definitions as it is a word that, applied in different social, historical, and cultural contexts, means different things. As a result, providing a simple definition is an effective way to distinguish the uniqueness of technology by trying to separate it from society while also taking into account how the social factors affect technological and scientific developments. Doing so, in turn, leads the way toward a more in-depth understanding of how technology intersects with other social realms. With this definition in mind, we continue our inquiry into how technology and society intersect.

The Intersection of Technology and Society

The study of technology has typically been approached from a material standpoint, consisting of the examination of tools and tool use. Early scholars of technology paid little attention to the social implications of technology, as illustrated in the definition of technology as material substance discussed earlier in this chapter. The focus on how the social and technological come together and influence each other started to become an object of study in the **Marxist tradition** around 1850 with its focus on **inequality**, which spurred interest in understanding how machines affect labour. For instance, the textile industry introduced machines to simplify and speed up work processes, resulting in the employment of non-skilled workers at lower wages (Berg, 1994); in the Marxist tradition, the focus was on how machinery was a central factor in deskilling and how it increased tensions in labour relationships. At this point in time, then, a transition occurred, away from the study of technology itself—as merely an object—to an interest in how technology changes social structure and brings about social change. We discuss this period in greater detail in Chapter 6 when we cover labour and technology.

Examining the social side of technology is essential as our society moves toward greater integration, what Ellul (1964) has described as a **technological society**. The concept of technological society does not describe technology merely as a tool that exists as an extension of our human faculties. Rather, it considers technology as part of a system that engulfs every aspect of human existence and in which no other human sphere is so central as that of the technological. Ellul (1989) coined the term "technological society" to emphasize that technology has a lasting impact on our environment and, therefore, on what we are as a society:

> [W]hat is at issue here is evaluating the danger of what might happen to our humanity in the present half-century, and distinguishing between what we want to keep and what we are ready to lose, between what we can welcome as legitimate human development and what we should reject with our last ounce of strength as **dehumanization**. I cannot think that choices of this kind are unimportant (p. 140).

We cannot ignore technology because it affects every aspect of human existence. On a typical day, we drive, talk via cellphones, email friends or co-workers, leave messages on Facebook, and rely on many of the other comforts provided by technology in the Western world. The far-reaching effect is such that rejecting technology, or being excluded from technological progress, has social, economic, and political consequences. The developing world falls behind in terms of opportunity, health, and standard of living simply because of the gap in technological sophistication. This will be a central theme of this book and can be described as **technological inequality**, which refers to the gap between the haves and the have-nots, resulting in disadvantages for the latter. We need to examine and question this inequality in order for society to be able to make smart choices about its technological future.

The problem with taking a narrow view of technology, technology as simply an object, is that it does not address the ongoing impact that technology has on society. Pacey (1983) sees technology as a complex system that includes people and machines. For him, this system exists in a historical, social, political, and cultural context. He argues that technology is not culturally neutral as it "is seen as a part of life, not something that can be kept in a separate compartment" (p. 3). When technology reaches the point of being embedded in the routines and practices of everyday life, it is said to have normalized.[3] (We will discuss the issue of neutrality of technology in greater detail in Chapter 11.) Feenberg also stresses the interplay between technology and the social: "[t]he introduction of a new technology into a given social environment is a powerful culture transforming act with immense political and economic consequences" (1982, pp. 17–18).

Box 1.1 discusses five challenges present in any inquiry of the relationship between technology and society. In addition, new technologies continue to emerge, creating new challenges for scholars. In the following section, simulation and augmentation are discussed as two recent technological developments that illustrate how our perspective of what technology is continues to evolve and also reveal the ethical, moral, and societal implications of these endeavours.

BOX 1.1 ⁑ CHALLENGES IN THE STUDY OF TECHNOLOGY AND SOCIETY

The study of technology and society presents numerous challenges. The first of these results from the complexity of the concepts themselves, as each is broad and difficult to define and measure. In addition, there are numerous challenges in how to approach research problems in the area. The five most prevalent challenges are as follows:[4]

1. *Rapid technological advances*: Technology evolves at a rapid pace. Not only do existing technologies transform, often becoming more complex, efficient, and fast, but there are also moments in history where technological revolutions have occurred. The key technological revolutions usually discussed include the discovery of fire; the invention of the wheel; the invention of the printing press; the discovery of electricity; and the dawn of the digital era. However, there is no definitive list of technological revolutions. We could include many other technological developments in our list, such as early tool use, navigation, metallurgy, functional magnetic resonance imaging (fMRI), etc. The point is that technologies evolve quickly and a single invention can lead to an infinite number of further developments.

2. *Unprecedented social change*: New technologies lead to unprecedented social change in many realms of society, affecting many aspects of social, cultural, and economic life, including how we work, live, communicate, and travel. Studying these changes is difficult because we cannot easily capture new forms of behaviour, thought, attitude, and belief with existing frameworks and methodologies. Thus, to assess the full impact of technology on society, researchers need to develop new theories and forms of measurement that complement existing ones.

3. *Direction and type of effect*: Determining the direction of effect is a major challenge because it is difficult to determine the extent to which technology affects society or society affects technology. Some analysts have argued that a mutual shaping of society and technology occurs (Bijker, Hughes, & Pinch, 1999), which is complex and non-linear (Chapter 3 covers this point in more detail). Moreover, researchers in the field aim to uncover an effect, regardless of whether the type of effect is positive or negative. However, in many cases no directional effect can be observed because technology becomes normalized and is then a part of everyday life.

4. *Target group*: When examining technology, we must keep in mind the target group. Many social changes associated with technology are specific to a particular social group, social class, ethnic group, or religious belief, which is the root of technological inequality. For example, industrialization

affected workers in different ways than it affected managers and owners. For workers, industrialization meant deskilling, lower wages, and inferior working conditions. By contrast, employers saw advantages in streamlining work processes, reducing error, and increasing efficiency. Further, the use of machinery allowed employers to hire unskilled labour for lower wages. Hence, the particulars of a group have to be examined to understand how that group is appropriating the technology and how the technology has an impact on members' lives.

5. *Changing uses*: The uses of technology also change over time. The cellphone, for example, started out as a portable device to communicate one-to-one independently of location. It quickly transformed into a more complex tool, allowing users to also access the Web, take pictures, listen to music, store information, and play games. Hence a single technology often mutates into a technology serving multiple purposes.

Contemporary Discussions of Technology

In addition to the definitions of technology we have discussed so far, there also exist contemporary conceptualizations that are directly influenced by our digital age. These contemporary definitions show how the evolution of technology leads to changes in how we define technology as well as to changes in the value and relevance we ascribe to technology in society. We discuss the two most prominent approaches here to show how radically different the notion of technology is in **simulation** and **augmentation**, where the material and the human start to blur and slowly become indistinguishable. Simulation is the attempt to develop technologies that can imitate the human mind and/or body. By contrast, augmentation is based on the integration of technology with the human body to enhance or strengthen certain capabilities or functions. These two perspectives question our view of technology as an external force because technology becomes either another actor in a social system or it becomes seamlessly integrated with the human mind and body.

Simulation

Artificial intelligence (AI) started as a discipline in the late 1940s, when computing was still in its infancy, with the aim of developing tools that could resemble or outperform human intelligence. While debate continues today as to what it means to have an intelligent machine, the general goal in AI is to simulate in any way or form human faculties, often including emotionality.

Alan Turing (1950) was the first to discuss the idea of a "thinking machine" and to propose ways in which we could test if machines can "think" or at least simulate thinking. His most central contribution to the field was the idea that AI could be achieved through computation instead of machinery. Early attempts to simulate human faculties had focused on building complex machines that physically resembled humans. However, according to Turing, what determines if a machine qualifies as intelligent is the extent to which it can imitate the capabilities of the human brain.

Turing (1950) argued that machines have the ability to imitate human cognitive processes to such an extent that it is impossible to distinguish between human and machine. He devised the **Turing test** or imitation game for two purposes: (1) to test a machine's potential to show intelligent behaviour and (2) to assess a human's ability to differentiate between human and machine intelligence. In the Turing test, three players—an interrogator, a human, and a machine—are visually separated from each other and the goal of the game is for the interrogator to discriminate between machine and human through a series of queries using natural language. If the interrogator cannot distinguish machine from human, the machine has successfully passed the test. The Turing test has had a profound impact on the field of artificial intelligence and computing in general, as it demonstrated the complexity of distinguishing what is profoundly human and what is in essence technological.

Even though computers can perform complex tasks mimicking humans or even outperforming them, John McCarthy (2007), one of the founders of the field of AI, argues that computers and humans are still fundamentally different in the way they solve problems: "Computer programs have plenty of speed and memory but their abilities correspond to the intellectual mechanisms that program designers understand well enough to put in programs" (para. 17). Hence computers can easily perform certain types of tasks—those based on algorithms—while other human abilities, even those that a two-year-old can easily master, cannot yet be programmed. Underlying this complexity is the lack of understanding in cognitive science about exactly what makes up human intelligence and how to program these abilities. In addition, McCarthy argues that a fundamental difference between humans and computers is that "[v]ery likely the organization of the intellectual mechanisms for AI can usefully be different from that in people" (para. 15).

Not surprisingly, the early dream of creating a machine that resembles humans has been largely abandoned even though it continues to persevere in science fiction. (Examples from science-fiction movies of machines that simulate human faculties and even resemble human appearance abound and include *A.I. Artificial Intelligence, The Terminator,* and *Sleeper*). Instead, much effort is now devoted toward building machines that can perform complex tasks that humans have difficulty with or are unable to perform because of limited memory and computational resources. These intelligent machines do not need to resemble humans in their appearance or in their emotionality.

One of the most successful recent developments in AI is that of **intelligent agents** or **chatterbots**, which are computational systems based on **algorithms** that can process complex and large data (Russell & Norvig, 2003). While there is much debate around whether Web-based agents can qualify as intelligent, some chatterbots have reached a high level of sophistication and usefulness. Joseph Weizenbaum in 1966 developed a prototype of a chatterbot, which he named ELIZA, that could convincingly simulate a Rogerian psychotherapist (Rubin, Chen, & Thorimbert, 2010). ELIZA was designed on the premise that patients direct the conversation, and as a result the chatterbot primarily relied on external input to hold a conversation (to interact with ELIZA, go to www.masswerk.at/elizabot/eliza.html).

More sophisticated bots are autonomous, that is, able to act without external input. An example is the chatterbot that greets visitors at the US Army website: **SGT STAR** is able to provide, in addition to basic information about the army, responses to more complex queries. To the question "Who are you?" SGT STAR responds: "My name is SGT STAR which stands for Strong, Trained and Ready. I'm an artificial intelligence agent created for the US Army to provide you with information about Army life." SGT STAR reflects the army's voice and has been programmed to provide responses that fit with the army's norms and values. After testing SGT STAR with a wide array of questions, it becomes apparent that he is a "machine" as some responses are quirky. Nonetheless, SGT STAR demonstrates advances in computing as he can act much more autonomously than ELIZA, who relies almost completely on external inputs to carry a conversation.

Modern AI systems show the current technological limitations in building intelligent machines and more fundamentally demonstrate that the question as to when a machine can be considered intelligent is very difficult to answer. Intelligent systems are becoming more prevalent as they can act autonomously without the need for human input.

"Chatterbot": SGT STAR, the US Army's virtual guide.
Source: © US Army. Used with permission.

BOX 1.2 ❖ AUGMENTATION THROUGH WEARABLE COMPUTING:
THE BIRTH OF THE CYBORG

The cyborg dream is no longer just science fiction. The first attempts to link human and machine occurred in the 1980s when Steve Mann started using a wearable computing system that he called the WearComp. This system integrates wires, sensors, and computers with the aim of increasing users' memory, enhancing their vision, and allowing them to stay perpetually connected to data (Mann & Niedzviecki, 2001). The cyborg state raises an intriguing question: What is it like to live as part machine and part human? Does doing so fundamentally change human nature?

The images below show the development of Mann's equipment from the early

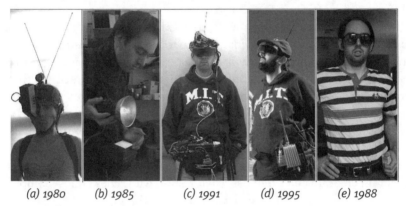

(a) 1980   (b) 1985   (c) 1991   (d) 1995   (e) 1988

Steve Mann's cyborg evolutions.
Source: Mann, S. (2012). Wearable Computing. In Soegaard, Mads and Dam, Rikke Friis (Ed.), *Encyclopedia of Human-Computer Interaction*. Denmark: The Interaction-Design.org Foundation. Available online at www.interaction-design.org/encyclopedia/wearable_computing.html. Used with permission.

1980s to the late 1990s. These early prototypes are in stark contrast to the current incarnation of his WearComp system, which is so minute that it is barely noticeable and consists of three main features. First, it includes a digital camera that allows Mann to upload a video stream of his daily experiences in real time to the Internet, giving new meaning to the concepts of privacy and surveillance (this will be discussed in more detail in Chapter 10). In addition, a small camera is attached to his eyeglasses, providing him with a split real-to-Web experience, connecting him to the Internet and mobile devices 24/7. Finally, the camera also allows him to block out stimuli from his environment, such as advertisements, presenting an edited image of his surrounding world.

As a result of the device, Mann experiences the world as an amalgamation of the physical environment that surrounds him and the digital world that streams in front of his eyes at all times through his Web interface (Mann & Niedzviecki, 2001). This divided existence shows that "Mann inhabits a different world than you or I, a 'cyborgspace' that he claims is inevitable for all of us" (Dewdney, 2001, D7). To what extent human and machine will be integrated is still an open question; however, the concept has already garnered many followers and the community of cyborgs has reached about 80,000 worldwide (Dewdney, 2001).

Augmentation

More recent discussions of technology have focused less on AI and more on ways to create augmented environments and bodies. Whereas AI is based on principles of simulation and replication between human and nonhuman entities, augmentation is based on principles of connectedness and responsiveness (Viseu, 2002). In augmentation, actors do not change their physical appearance and body shape, as is the case in simulation. In this approach "[r]ather than building self-contained machines or leaving the body behind, machines and humans are coupled together into new hybrid actors with added capabilities" (Viseu, 2002). The physical body is augmented by connecting it to digital components with computational and communicational capabilities.

Augmentation can occur in many ways. An example of augmentation is **wearable computing**, where portable computers aid in the decision-making process by providing additional information, enhanced **surrounding awareness capabilities**, and real-time data streaming. A cellphone is a wearable computing device because it augments our capacities to communicate by bridging space and time constraints and allows us to access data on demand (Klemens, 2010).

While the aim of artificial intelligence is to *recreate* the human experience, augmentation attempts to *intensify* human qualities. One such example of augmentation is the **cyborg**, a merger of human and machine through seamless connectivity. Clynes and Kline introduced the term *cyborg* in the 1960s to describe the ways that long-distance space travel would require astronauts to transform their human qualities to robot-like qualities, which are fully automated. These transformations would allow astronauts to master the challenges presented in extraterrestrial environments. For Clynes and Kline (1960), the main objective of the cyborg was "to provide an organizational system in which such robot-like problems are taken care of automatically and unconsciously, leaving man free to explore, to create, to think, and to feel" (p. 27). Science-fiction writers then adapted the cyborg concept, which became a part of our popular culture as presented in books, movies, and television programs. A cyborg—a robot-like machine with human characteristics—is featured in the movie series The Terminator, in which Arnold Schwarzenegger plays Terminator, who has super-human strength and is indestructible by contemporary weapons. The latest Terminator model, the T1000, is superior to its predecessor model in many ways. It consists of a mimetic polyalloy, which is a kind of liquid metal that can emulate anything that surrounds it, giving the cyborg enormous flexibility and superiority. Interestingly, the cyborg also displays emotions that allow it to blend in well with humans, unlike in *Terminator*, which was devoid of any human-like emotions.[5] Box 1.2 describes the cyborg experience as an example of an artifact built to augment reality through the merging of human and machine.

Significantly, the cyborg concept is moving from the annals of science fiction into the mainstream with the development of a number of man–machine systems that are geared toward practical goals, such as dealing with impending medical conditions. Such a prototype is the **C-Leg** system, which allows individuals with an amputated leg to walk again.

Futurists predict that augmentation and the coming together of human and machine is inevitable. One of the more controversial possibilities is **transhumanism**, which is characterized by the "surpassing of the biological limitations of our bodies, be they our lifespans or the capabilities of our brains" (Dewdney, 1998). Transhumanism is the product of a quasi-religious movement dedicated to the belief in the "futuristic technological change of human nature for the achievement of certain goals, such as freedom from suffering and from bodily and material constraints, immortality and 'super-intelligence'" (Schummer, 2006, p. 430). According to Schummer (2006), transhumanists have "an existential interest in nanotechnology, as a means for the ends of personal and/or societal Salvation" (p. 432). The transhumanist position toward nanotechnology raises pertinent ethical questions about the role of technology in transforming and prolonging the lifespan of the human body. Moreover, transhumanism aims to use scientific discoveries to enhance human faculties. The Kurzweil Accelerating Intelligence blog (www.kurzweilai.net/) showcases recent technological discoveries and inventions that have the potential to radically transform the human body. For instance, findings are reported from a study that shows how mice learn faster and remember better after PKR (an immune molecule that signals viral infections to the brain) is inhibited. From a transhumanist perspective, PKR is seen as a potential smart drug that boosts our cognitive capacities.

Developments such as the cyborg C-Leg and the inhibition of PKR illustrate the advancements in augmentation and the ways that science and technology can become a part of the human body, making it increasingly difficult to trace the boundaries between what is human and what is technological.

Conclusions

With advancements in science and technology, the level of interaction between society and technology continues to grow rapidly. The prevailing theoretical perspectives on this subject have supplied a rich foundation of debate and criticism in this area. Yet, as the technology–society relationship deepens in complexity and intimacy, our definition of technology also becomes more and more opaque. In Western societies, the dominance of technology in everyday life has greatly altered conceptions of work and play. Yet, such dominance has also profoundly affected and shaped notions of human existence in ways that earlier critics could not have imagined. Whereas early opponents of industrial technology were suspicious of hulking steel machines displacing the skilled craftsman, contemporary critics are assailed with new ethical and moral challenges, as infinitesimal technologies enter the economic marketplace and the politico-cultural spheres of our secular, consumerist society.

The philosophical implications of classifying and clarifying present and forthcoming technologies will arguably be intertwined with how contemporary and future generations of citizens and scholars view technology as a whole. Indeed, it will be interesting to see how members of Generation Z (those born between 1990 and 2010) will rationalize, critique, and characterize technology, especially when considering the indelible role of, and their dependence upon, technology in their lives.

Questions for Critical Thought

1. Discuss the challenges associated with studying technology. Choose a specific technology of your choice to illustrate the discussed challenges.

2. How is technology described in the definition of technology as practice? What are the strengths and limitations of this conceptualization?

3. Compare and contrast the theoretical view of technology as power with that of technology as mechanization.

4. Critically examine the concept of the cyborg and discuss its ethical implications.

Suggested Readings

Ellul, J. (1964). *The technological society*. New York: Knopf. The book provides a critical analysis of the integration of technology and society by examining and contrasting various theoretical understandings of technology.

Franklin, U.M. (1992). *The real world of technology*. Concord, ON: House of Anansi. This book critically examines the relation of society to technology through the framework of practice.

Mann, S., & Niedzviecki, H. (2001). *Cyborg: Digital destiny and human possibility in the age of the wearable computer*. Toronto: Doubleday. An exploration of the wearable computer and how it affects humans and society.

Rubin, V.L., Chen, Y., & Thorimbert, L.M. (2010). Artificially intelligent conversational agents in libraries. *Library Hi Tech, 28*(4), 496–522. This review paper examines the current use of artificial intelligent systems in libraries.

Online Resources

Steve Mann's laboratory at the University of Toronto
www.eecg.toronto.edu/~mann/
The webpage compiles papers, presentations, and video footage of Mann's research about technology, privacy, and digital tools. It also includes information on his book and the movie made about his life as a cyborg.

What Is Artificial Intelligence?
www-formal.stanford.edu/jmc/whatisai/whatisai.html
John McCarthy, the founder of artificial intelligence, provides a brief introduction to the topic.

Ask SGT STAR
www.goarmy.com/ask-sgt-star.html
This is the intelligent agent of the US Army, which answers questions about army life for those visiting the website.

2 Technology in Society: A Historical Overview

Learning Objectives

- To obtain a broad overview of the history of technology.
- To critically examine the mutual shaping of technology and society in a historical context.
- To provide an alternative view of the history of technology by taking a sociological perspective of technological transformations.
- To learn about how digital tools have led to the development of a high-tech society characterized by customization, individualism, and privatization.

Introduction

This chapter provides a broad overview of the history of technology by tracing the roots of technological development, discussing key periods of technological innovation, and outlining our present-day, high-tech society. The primary aim of the chapter is not to comprehensively cover technological developments over time but, rather, to outline on-the-ground examples that illustrate how technology and society intersect. Because technology is the strongest force of change in society, its development, transformation, and diffusion directly shape many aspects of society, such as work, community, and social relationships. This chapter further attempts to link technologies, inventors, and historical moments to provide an in-depth examination of the socio-political context from which technologies emerge. This analysis will illustrate how we need to understand technology as a complex process instead of viewing it as a purely technological entity disembodied from any societal context. This chapter shows how technology is ingrained in society, affecting all aspects of our lives, and it outlines the merits of taking a sociological perspective of the history of technology.

Why Study the History of Technology?

Before we embark on historical details, we need to reflect on why it is important for us to understand the history of technology in our society. Technology is seen as a key mechanism of social change, moving society forward into new social, cultural, political, and economic milieus. For example, it is technology that makes it possible for humans to subsist in **megacities**, such as Tokyo, with a population of 33 million, or Mexico City, with a population of 22 million. But megacities did not emerge overnight; in fact, there is no evidence of human settlements in pre-historic times. The creation of megacities was possible only through the constant advancement of technology over centuries. This process occurred simultaneously with the evolution of advanced forms of construction, chains of mass distribution of goods, and new forms of social and political organization. While the evolutionary process of technology was not always straightforward, understanding that process and its link to social change is important.

Scholars of the history of technology describe the history of tool use as following an **evolutionary model of technological development** because tools build upon existing knowledge, making their development incremental. In history, inventors do not work in complete isolation but continue to draw upon earlier designs, often combining existing pieces into new forms. To better understand how this evolutionary model unfolds, George Basalla (1988), in *The Evolution of Technology,* compares the process of innovation to the way biological evolution occurs in the natural world. The comparison is based on the diversity of tools that exists in the man-made world, which he argues is similar to that of life forms that exist in nature. Central to Basalla's argument is the idea that novel tools arise only from earlier ones and "that new kinds of made things are never pure creations of theory, ingenuity or fancy" (Basalla, 1988, pp. vii–viii). Eugene Ferguson (1992) uses the example of the automobile to illustrate how technology is evolutionary instead of revolutionary, as the main components constituting the automobile were based on existing inventions. The pistons, cylinders, and nozzles already formed part of a pump designed by Hero of Alexandria two millennia ago. What made the automobile different from previous technologies was its unique blend of techniques, including modern electrical components. This example shows that the integration of existing ideas in new ways is often what makes a tool innovative.

That technology develops in an evolutionary way, rather than in a revolutionary manner, may suggest that the study of technology is straightforward. Nonetheless, two key issues make understanding the history of technology difficult. First, there is a gap in time that exists between when an inventor first develops a new technology and when this technology is revealed and examined. This gap represents a time warp because it makes the past inaccessible. Because social, cultural, and technological transformations are so extensive, people find it basically impossible to relate back to previous eras. And the further back in time one travels, the more difficult it is to understand human behaviour. The **Stone Age**, for example, occurred approximately 2.6 million years ago, while the medieval period was only approximately 500 years ago. If we find it challenging today to relate histori-cally to the medieval period, it is even more difficult to study events dating back millions of years.

The second issue that makes it difficult for us to understand the history of technology is our reliance on artifacts as evidence. Artifacts can be con-sidered the basic components of **hominin material culture** and are usu-ally contrasted with elements encountered in nature, such as leaves, trees, caves, etc. *Material culture* refers to the interrelation that exists between an artifact and the social relations, cultural attitudes, and norms that are pres-ent in the society that uses the artifact.

Using artifacts as sole evidence of how societies evolved, however, is limit-ing because identifying all the existing tools of the time period under study is a daunting task. Morever, studying the artifacts alone does not take into account how they shaped social norms, attitudes, and behaviours. The first tools were primitive and consisted of bones, flint, stones, and sticks, making it difficult for scholars to distinguish from naturally occurring objects. In order to make the past accessible, then, scholars study how previous societ-ies used artifacts and what their social significance was. Box 2.1 describes a number of techniques that have been proposed to study artifacts and discusses their strengths and weaknesses.

Box 2.1 ❈ **Digging Technology: Tool Use in the Stone Age**

Physical anthropologist Thomas Plummer (2004) distinguishes between two key techniques for discerning technological transformations: (1) **excavation** and (2) **actualistic studies**. Excavation is the predominant method of investigation in archaeology and refers to the systematic exposure, preserving, and recording of artifacts from archaeological sites, or **digs**. Specialists employ a number of excavation techniques:

- **Open area excavation**: In this technique, excavators dig large trenches of at least 10 metres by 10 metres, sometimes even several hectares, with no predefined balks or sections.

- **Planum method**: In this method, excavators systematically remove arbi-trarily defined slices or spits of even thickness to uncover artifacts. The advantage is that artifacts can be directly assigned to the respective spit and further divided spatially with the help of a grid.

- **Quadrant method**: This method is particularly suited for circular sites, such as roundhouses or barrows, and consists of dividing the site into four equal quadrants for examination.

- **Wheeler method**: Sir Mortimer Wheeler developed this technique at sites in India and southern Britain. In this method, excavators dig trenches in the shape of rectangular boxes with balks between to allow the excavators to systematically record and analyze finds in each trench.

In addition to conventional archaeological explorations, actualistic or experimental studies have emerged as an alternative technique for investigating the validity of the **paleoanthropological record**. Researchers use innovative and often practically oriented techniques. For example, these researchers replicate and test artifacts to determine what kinds of behaviours they facilitated and the kinds of changes in the environment their use may have caused. Researchers have also conducted extensive research examining the factors determining the movement of resources, site formation, and food-sharing among modern hunter-gatherers. The advantage of actualistic studies is that instead of relying only on the archaeological evidence, researchers test the extent to which the artifacts support a specific way of life. Therefore, the goal of actualistic studies is to recreate past ecosystems in order to replicate the conditions in which hunter-gatherers lived. Anthropologists involved in actualistic studies may, for example, live for a short period of time in a manner similar to hunter-gatherers in prehistoric times, attempting to reproduce the experience as closely as possible.

Each of the techniques discussed in Box 2.1 has been used in different situations to study artifacts from the past. Research, including that done by anthropologists using these techniques, has determined a series of technological periods in human history. Next, we start our examination of the history of technology with an overview of these different technological periods and the associated socio-political and cultural contexts. We define six distinct periods of technological evolution: (1) the early beginnings of technological ingenuity, (2) ancient technology, (3) the Renaissance, (4) the Industrial Revolution, (5) electronic times, and (6) the information society.

The Stone Age: The Early Beginnings of Technological Ingenuity

The earliest evidence of technology is traced back to the use of tools during the Stone Age, about 2.6 million years ago. There are two reasons why the study of early tool use in prehistoric times is significant. First, tool use suggests that humans reached an awareness of themselves as separate from the world around them, allowing for the development and use of external objects to accomplish tasks. This represents the earliest manifestation of human behaviour as intelligent and goal-oriented. Second, tool use is linked to human social and cultural evolution in terms of setting the foundation for a wide range of task-oriented behaviours, such as cutting, scratching, and manipulating (Plummer, 2004). These kinds of behaviours, even though simple, allow for complex interactions with the environment and other individuals and are the early beginnings of consciousness and thought. A set of socio-cultural behaviours directly results from tool use—users feel a sense of ownership over objects and, as a result, protect these from potential intruders. Therefore, even early tool use necessitates a sociological and cultural understanding of the relationship between tools, humans, and social systems.

In the Stone Age, communities would subsist on meat, by hunting prey, and produce, by gathering berries and other edible plants; this is how the term **hunter-gatherer societies** was coined. Hunter-gatherers lived in **nomadic tribes** that were highly dependent on climate and food availability. During this period, stone tools started to emerge as the first key technological development. Plummer (2004) describes how Oldowan archaeological sites in Eastern, Southern, and Northern Africa provided first concrete evidence of material culture. Beginning around 2.6 million years ago, stone cores were struck to make sharp-edged chips of stone known as flakes. Tool use at the time was limited to animal butchery and possibly preparing plant foods and working with wood. These materials developed from single all-purpose tools to an assemblage of varied and highly specified ones, demonstrating the high level of complexity involved in early tool development. Though there is an assumption that the hunter-gatherer lifestyle was deprived, lonely, and brutish, Late Stone Age societies had a rich social and spiritual life—as evidenced in cave paintings, rock engravings, and other existing rituals.[1]

Both agriculture and the domestication of animals are considered technological inventions. The introduction of agriculture marked a radical departure from the hunter-gatherer lifestyle in that settlements started to develop around food production and consumption. In addition to domesticating animals, former hunter-gatherers also began to follow herding patterns, either seasonally or continuously, and to engage in animal trade with other tribes. This pastoral shift is usually referred to as the **Neolithic period**, where the Greek word *neo* refers to "new" and *lithic* refers to "stone," that is, the New Stone Age.

The most influential scholar of this period is the archaeologist V. Gordon Childe (Ratnagar, 2001). After careful investigation of the archaeological record in the Middle East, he concluded that the move from hunter-gatherer societies to the reliance on tools for agriculture was the most significant change in human cultural and social evolution. To denote the impact of the introduction of agriculture on human societies, Childe introduced the term **Neolithic Revolution** (often also referred to as the Agricultural

Revolution). For sociological inquiry, this is a turning point because settlements allow for the study of place as a social phenomenon, and lead to the introduction of sociological concepts such as identity, sense of belonging, **community**, and social structure. In **sedentary societies**, individuals are assigned different roles, with some working in agriculture, others continuing to hunt and gather, and yet others taking care of the settlements with food preparation and other tasks. The Neolithic Revolution marked the emergence of social roles, class, and work relations.

Christopher Boehm (1993; 1999; 2000) describes Neolithic societies as having fairly egalitarian structures as a result of, among other factors, technologies such as agriculture and animal domestication, which provided sufficient nourishment for the existing settlements. This sets the beginning for more complex sociological analysis with social structure emerging as a means for distinguishing people in society on the basis of labour, labour relationships, and the emergence of power struggles between groups of people occupying different social roles.

Ancient Technology: The Development of the Scientific Method

As technologies such as the smelting of iron, the screw, and the lever emerged in Greece around 500 to 336 BCE and in Rome around 509 BCE to 190 CE, social organization changed radically. Technology facilitated certain forms of life—concentration in cities and improved food production—and helped in the creation of complex government structures. While Greece and Rome basically continued threads of technological progress that had begun in earlier times, the most important contribution to science and technology from that time period comes from the premise that thought forms the basis for practical action.

Indeed, Fowler (1962) starts his book *The Development of Scientific Method* with a comprehensive overview of ancient Greek philosophy, where he illustrates that it was the ideas and thought processes of the time that represented the early beginnings of technological innovations—not the tools themselves. Then, once the appropriate knowledge was in place, technologies developed as practical solutions to everyday problems. While ancient civilization may be known for its great thinkers, not all inquiry was abstract; in fact, certain ideas also had practical significance. For example, the study of **perspective effects**, which refers to an array of techniques designed to provide the impression of depth and symmetry, had an impact on building construction, first in Greece and then in Rome. Moreover, the Romans were innovators in building technology by using materials such as fired brick, tile, and stone. To shore up their construction, they also invented a new type of cement that was strong and also set under water. In their buildings the Romans designed architectural features, such as the arch, the vault, and the dome, each adding considerably to the possibilities in construction. These techniques allowed them to build large structures such as amphitheatres, aqueducts, and tunnels, some of which still exist today. The amphitheatres allowed for social interaction and represented some of the first forms of mass entertainment. These developments had enormous social significance as they allowed for new forms of social organization to emerge, such as the gathering of townspeople in social spaces for debate and entertainment, the development of more densely populated cities, and the emergence of increased traffic and trade.

Many of the ideas around science and technology that were formulated in ancient Greek philosophy continue to be present in the twenty-first century. This is particularly true in the case of **basic research**, the idea of scientific inquiry not as specific practical applications but, rather, as the development of knowledge about the world. The Perimeter Institute for Theoretical Physics in Waterloo, Canada, exemplifies this emphasis on basic research, with about half a billion dollars of public and private funding invested into examining the fundamentals of physics.

Applying the scientific method of inquiry to the study of social processes is not new but, rather, emerged in ancient Greece and was further developed during the Roman era. The tension between basic and applied research continues to exist in the twenty-first century and is visible both in society's approach to funding research and in the mandate of universities, which seek to balance the need for technological innovations with programs geared

The University of Waterloo's Perimeter Institute for Theoretical Physics is dedicated to studying fundamentals about the world around us and sparking innovation. It attracts international scholars with its dynamic research atmosphere and it conducts outreach programs for youth, teachers, and the general public. Its architecture—pictured here is part of its atypical, articulated facade—reflects its forward-thinking principles.
Source: © iStockphoto.com/Sebastian Santa.

toward simply acquiring and diffusing new knowledge that has no direct economic impact.

The Renaissance: The Awakening of the Mind through Technology

Historians often characterize the invention of Johannes Gutenberg's printing press in the mid-fifteenth century as groundbreaking. As a tool, its use reshaped history and helped bring about early elements of modernity, including mass production, mass distribution, and the notion of a mass audience. The printing press has also been linked to the spread of literacy and to the diffusion of new ideas throughout Europe, which were crucial in promoting the development of novel religious, scientific, nationalist, and secular thought. Indeed, the printing press proved to be an instrument of great political, economic, social, and cultural power by (1) enabling the spread of ideas, which circumvented traditional institutional bodies such as the Roman Catholic Church; (2) codifying language; and (3) bringing about the commoditization of information.

Before the development of Gutenberg's printing press and prior to the onset of print culture, manuscript or scribal culture was the primary source of written texts. In workshops, trained scribes would carefully and painstakingly copy texts to produce hand-copied books (Edmunds, 1991). These books, often of a religious nature, were expensive and time-consuming to produce due to the measured and methodical nature of scribes. As a result, books were accessible only to a select group of people who were literate and affluent, and could read Latin and/or Greek.

Gutenberg's printing press combined his own innovations with pre-existing technologies into a new tool. Prior to Gutenberg, woodblock printing, which had originated in China, was the most popular form of printing in Europe. However, the methods originating in China lacked uniform standardization, requiring skilled craftspeople to develop individual wooden blocks. Inferior or worn blocks often distorted lettering and produced errors (Eisenstein, 1979). Gutenberg, who was a goldsmith from the German city of Mainz, developed Europe's first movable metal type by applying his knowledge of metallurgy—a process that had originated in Korea that same century (Füssel, 2005). Gutenberg split the text into individual components, such as letters and punctuation marks, which were cast in metal and assembled to form specific words. This yet again is evidence of how techniques develop in an evolutionary manner rather than following a revolutionary pattern, borrowing from existing techniques and adding new elements to yield new forms of production.

The printing press led to an era in which oral culture gave way to print culture. Marshall McLuhan proposed that the development of the printing press coincided with a change in how information was transmitted and received: from oral transmissions that relied upon "the magical world of the ear" to processing information from textual communications that emphasized visual cognition (1962, p. 18). Elizabeth Eisenstein (1979), one of the most influential scholars who studied the effects of print on society, agreed with McLuhan; she also saw print as revolutionary and as changing society. She argued that the shift from script to print "created conditions that favored new combinations of old ideas at first and then, later on, the creation of entirely new systems of thought" (p. 75).

The development of the printing press brought with it changes in how texts were used, distributed, and marketed (1979), once again demonstrating the link between technology and society. Over time, the individualistic, producer-oriented approach of manuscript culture yielded to a consumer-oriented typographical print culture. This enabled the production of greater quantities of a single item within a much shorter time frame. Whereas manuscripts were primarily produced according to the literary interests of a single individual or institution, print texts could be manufactured according to the broad and diverse interests of a wider audience. The printing press altered the economics of the book industry by lowering the cost of purchasing books, which allowed individuals to assemble personal libraries, accrue new ideas and educate themselves about subjects of interest. Although not entirely error-free, the printing press brought about increased uniformity and encouraged the standardization of languages, as well as a surge in texts written in the vernacular (Febvre & Martin, 1997; McLuhan, 1962).

Without the printing press, it is arguable whether or not the dissemination of information via controversial texts and pamphlets during historical movements, such as the Reformation, the Enlightenment, and the Scientific Revolution, could have taken place. Some scholars, such as Haas (1996) and Johns (1998), have suggested that, as mentioned earlier, the printing press was more evolutionary than revolutionary: the product of a series of cultural, social, and technological changes occurring in European society during this period, whose influences were gradual rather than immediate (Haas, 1996; Johns, 1998). Others have noted that the printing press's effectiveness was largely restricted to areas of Western Europe (remember that China had been using movable type 500 years prior to Gutenberg's printing press) (Lorimer, Gasher, & Skinner, 2010). Regions consisting of largely oral-based cultures or that used non-phonetic alphabets in fact eluded the socio-cultural and political upheavals typically associated with the printing press in Western Europe (Briggs & Burke, 2009).

In conclusion, it is likely that the spread of new ideas about science, politics, and religion in Western Europe was the product of technological diffusion.

The cultural, political, and economic conditions of mid-fifteenth-century Western Europe were ripe to absorb and transform this technology into a powerful instrument of change. In addition, the printing press allowed new ideas about technological developments to spread more rapidly, leading to enormous advances in science and technology. We discuss in the following section how the Industrial Revolution—with the introduction of machinery into the workplace—changed the labour conditions of workers throughout Europe. Analysis of the Industrial Revolution demonstrates how the introduction of technological innovations can not only disrupt the social structure but also fundamentally transform all aspects of society, from labour relations to the very entity of the family.

The Industrial Revolution: Revolting against Technology

The **Industrial Revolution** (1712 to 1910) advanced technology in three key industries: textile, steam, and steel. The era of industrialization was significant for changing the production and distribution of goods and had a major impact on workers as well as consumers. The era brought with it radical changes in people's work processes, man–machine interaction, and worker–employer relationships. The introduction of machines coincided with an increase in the amount of individuals being employed in sectors, such as weaving in the textile industry (Berg, 1994). The enlarged pool of workers, consisting of an "unregulated" workforce of women's and children's labour, coupled with simplified technological processes associated with new machinery drove down wages and led to an increase in unemployed yet highly skilled craftspeople (Berg, 1994).

A new phenomenon also emerged during this time. Machines were no longer seen as neutral but, rather, were perceived as "actors" within a complex socio-technical system. In this system, technology was identified as the key force in producing social disparity and disrupting work conditions. To retaliate, workers geared their anger toward the machines that they perceived were taking away their jobs; the term **machine breaking** was introduced to describe workers' destruction of technology (Hobsbawn, 1952). These acts were organized by workers because they viewed the machines as threats to standards of living, employment, wages, and working conditions (Dinwiddy, 1979; Hobsbawn, 1952). Hobsbawn (1952) identifies two primary motivations for machine breaking: (1) to put pressure on employers to improve wages and working conditions, and (2) to direct hostility toward the machines themselves, especially those perceived as labour-saving. For worker groups without a tradition of trade or union organization, "machine breaking and violence proved effective methods of protest" (Randall, 1991, p. 149).

The Industrial Revolution was an era of unprecedented social change. Much of the upheaval taking place during this era was linked to the introduction of technological innovations in Europe. However, technologies alone do not lead to social change; we need to study their effect on society as part of complex social systems (for more details see Chapter 3). That is, it was not technology that led to craftspeople losing their jobs, as machine breakers at the time concluded. It was, rather, the interplay between technology and the existing social system that radically transformed social structure. From a historical standpoint, industrialization was only the beginning of large-scale change linked to technological innovation. Next, we discuss how innovations in mass media, such as cinema and television, transformed the flow of information in society, created new forms of mass cultural production, and introduced new notions of entertainment and propaganda.

Electronic Times: Hot and Cool Media

The Industrial Revolution was a time of upheaval as new technologies infiltrated the workplace, creating major disruptions for workers. The pace of technological development did not halt with the industrial era, however. Major changes were observed in particular in the late nineteenth and early twentieth centuries in what we refer to as the mass media or the electronic media. The widespread use of mass media and their impact on daily life have only recently become topics of much public debate and scholarly attention. Earlier, mass media had become such a part of daily life that users did not reflect much upon their effects on family life, on civic engagement, on the diffusion of information, and on community. Marshall McLuhan (1962; 1964) was among the first to draw attention to the effects mass media had on society, commenting on how these tools shaped our thinking processes as well as our understanding of the world. At the time, mass media encompassed radio, cinema, and television, and together they created a new era of electronic communication.

In order to understand the close link between mass media and society, McLuhan used provocative, bold aphorisms that he termed **probes**. His probes are ideas that challenge us to think more deeply about the link between messages, media, and audiences, and are not to be taken literally but, rather, metaphorically. Perhaps McLuhan's most well-known probe is **the medium is the message**, which expresses the idea of media having a direct effect on how people make sense of information. McLuhan (1962; 1964) argued that media have different characteristics, thus engaging the audience in different ways. The essence of "the medium is the message" is that most analysis of the effect of media on society focuses on content. For example, while Albert Bandura's (1973) well-known experiments examined how children learn aggressive behaviour from exposure to images on a screen, McLuhan argued that attention to the effects of the medium itself was often neglected. For McLuhan, then, the medium was as important as the content, if not more important, because it framed how society interacted with information. This understanding of the effect of media led McLuhan toward a deep exploration of both the relationship between media and society as well as the effect that different media have on society. To exemplify this link between media and society, Box 2.2 examines the way cinema was used in the early twentieth century as a means of propaganda.

BOX 2.2 ❖ CINEMA AS A PROPAGANDA TOOL IN SOCIETY

Originating in the late nineteenth century, the cinema became a mass medium that attracted people to the moving image. Film can be described as a technology-based art form that creatively expresses the world around us and that can be utilized to influence political or social change. Roebuck notes "the importance of the cinema to social development lay in its national coverage and the enormous size of its audience" (1982, p. 119). Soon, cinema's ubiquitous presence became "a great standardizing agent which brought huge numbers of people together through vicariously shared experiences and promoted a general worship of the same heroes and the formation of similar dreams" (Roebuck, 1982, p. 119). Early in the medium's history, films were also used to express views supporting a specific ideology, such as the films of Sergei Eisenstein in Soviet Russia, or a revisionist take on history as exemplified by D.W. Griffith's 1915 film *The Birth of a Nation*.

The intricate connection between cinema and society can be demonstrated in the case of Nazi Germany. In an era before the widespread popularization of television, the cinema performed a crucial role in providing people with visual slices of information and entertainment through newsreels and feature films. Under Nazism, the German film industry was controlled through the Reichsfilmkammer, which was governed by the Ministry of Propaganda, led by Joseph Goebbels. Films could only be created and used in accordance with the guidelines and objectives set by the Nazi party. A "popular medium and vehicle of mass culture, film preserved old forms of identity while offering a new (and powerful) instrument of consensus-building" (Rentschler, 1996, p. xi). Hitler argued that film and other similar media "could be instrumental in winning followers and propagating Nationalist Socialist ideas" (Hake, 2002, p. 77), and, at the same time, Goebbels "sought to transform the film apparatus into a tool of the state, a medium of myth, legend, and fantasy which would relieve reality of its dialectical complexity" (Rentschler, 1996, pp. 67–68).

As Minister of Propaganda, Goebbels wanted German films to "conceal the intentions" of the works by emphasizing presentation rather than content (Hake, 2002). As a result, the films produced under Goebbels aimed to "convey their messages as inconspicuously as possible" (Trimborn, 2007, p. 79). Even though some films were openly propagandistic, "most popular film genres carefully avoided references to the regime" yet continued to serve "it by promoting the sexist, nationalist and racist ideologies essential to its existence" (Hake, 2002, p. 77). This was accomplished by using the seductive language of cinema to create emotive films designed to collectively connect people to the shared experiences of the films' clichéd characters, as well as to the traditional values and illusory fantasies expressed in these motion-picture spectacles. The cinema thus became a mass medium adeptly utilized to placate or arouse the emotions of the masses.

Similar to cinema, television is an electrical device that provides information and entertainment to a large audience. By 1970, many North Americans had a television set in their homes and the average person was watching TV nearly four hours a day (Bogart, 1972). Nielsen data show that in 2009 Americans continued to spent large amounts of time watching TV in the home, including time-shifted TV, Internet TV, online videos, and TV via cellphones. In fact, the average American watches about 153 hours every month, a 1.2 per cent increase from 2008 (Nielsen, 2009).

McLuhan (1962; 1964) saw television as a unifying force that provides viewers with a sense of connectivity. To describe the social connections created by television, he introduced the notion of the **global village**. The global village represents the possibility that electronic forms of communication and media, such as television, can compress the rigours of spatial distance by enabling people to remain connected to activities going on anywhere in the world. McLuhan explains in an interview that "[t]he global-village conditions being forged by the electric technology stimulate more discontinuity and diversity and division than the old mechanical, standardized society; in fact, the global village makes maximum disagreement and creative dialog inevitable" (E. McLuhan & Zingrone, 1995, p. 259).

There are additional key terms to understanding electronic times. For example, McLuhan distinguished between **hot** and **cool media** or "cold" media. For him, media could be differentiated in terms of the degree to which they engage a consumer. Books were categorized as being "hot" because they engage a single sense, in this case the visual system, providing readers with large amounts of information that is processed at lower sensory levels.[2] By contrast, television was determined as being "cool" because it requires greater effort to determine meaning. In addition to the cool–hot dimension, McLuhan also distinguished between **high-** and **low-definition media**. Crisply detailed and well-defined information, such as detailed images, that required little effort to be processed were referred to as "high definition" and were found in hot media. By contrast, "low-definition media," such as cartoons, provided users with less information and thus required a greater participation from the senses in order to be understood.

An implication of McLuhan's theories is that over time existing and emerging forms of media transform personalized experiences and societal structures. Previously popular media formats, for example, often give way to new technologies, which repackage older forms of content in a manner responsive and applicable to the changing information demands and experiences of their audience. For instance, contemporary communication technologies and media, such as cellphones or social networking sites, offer users a greater degree of mobility, reach, and instantaneity than their predecessors (Klemens, 2010). As a result, these new devices have changed social modes of interaction and societal expectations for how communication can and should be conducted in the twenty-first century.

This section outlined the significance of electronic media or mass media in the history of technology. The developments in mass media that occurred at the end of the nineteenth century and the first half of the twentieth century transformed the flow of information in society, created new forms of mass cultural production, and introduced new notions of entertainment and propaganda. Clearly, the cinema became a new means of reaching mass audiences and served not only as a form of entertainment but also for political propaganda. Some analysts stipulate that mass-produced cultural goods lead to standardization and uniformity in content, and additionally destroy individuality and multiplicity of choice (Gasher, Skinner, & Lorimer, 2012). In their view, exposure to standardized cultural goods—for example, movies produced in Hollywood—caused members of society to become a homogenous, uncritical, and passive mass with little willpower to resist the appeal and influence of the mass media. While some of these extreme views have been dismissed, there is no doubt that these technologies have had a large-scale impact on the production, dissemination, and interpretation of cultural goods. Perhaps the most fundamental social change is the link between mass media production and consumption and the emergence of a mass audience at a scale that did not exist prior to the media of mass communication.

The Information Society: The Bits and Bytes Revolution

What does it entail to live in a high-tech society? To what extent do technologies impact who we are as individuals, as nations, and as a society? Sociological work has extensively scrutinized the effect of industrialization and urbanization on cities, communities, family structures, and work. To illustrate the kinds of societal changes brought about by current technologies (Haigh, 2011), we discuss here the iPod and the cellphone. We examine the iPod as a form of entertainment that has become a force of individualization, privatization, and customization among youth. It allows young people to feel empowered and to make individualized decisions about music choices in a private space away from the adult world. By contrast, we examine the cellphone in reference to the have-nots, to those in society who only marginally interact with technology but yet are still struck by it.

iPod Culture

Released in October 2001, the iPod ushered in a new era of portable media devices and quickly transformed Apple from a niche computer manufacturer into an international media provider (Hartley, 2009). By 2008, Apple was the world's biggest music retailer and had 70 per cent of the market share in digital music-player devices (Hartley, 2009). Initially designed to allow users to carry their favourite songs on a small compact transportable unit, the iPod is an example of a product that evokes the principles generally associated with twenty-first-century technologies of the information society, including mobility, speed, multimedia interactivity, ease of use, customization, and multi-purpose usage.

With its signature aesthetic appearance, evidenced in its click wheel and white earbuds, the iconic iPod is a clearly identifiable product that along with its off-shoots—the iPod Shuffle and the iPod Touch—has transformed the manner in which media is consumed, transmitted, transported, and received. Part of the iPod's initial popularity can be traced back to changes within the music and telecommunications industry. In 1993, the MP3 file format became popular for storing digital audio due to its relatively small size and versatility. Simultaneously, peer-to-peer networks and BitTorrent sites, such as Napster, Kazaa, and The Pirate Bay, provided Internet users with the ability to (often illegally) download music files. Subsequently, the iPod became a convenient device that allowed users to take their music collection with them.

iPods have brought about new forms of customization, which can reflect their users' unique personalities. The diverse selection of media content stored on the device is in itself reflective of individuals' interests, from the soundtrack collection to their choice of podcasts and applications (a.k.a. "apps"). Additionally, the exterior of the device can be altered according to the respective user's tastes by purchasing an iPod featuring a favourite colour, by adding a personalized engraving, or by adding stick-on artwork known as "skins."

The new forms of control and customization that iPods render go beyond the look and feel of the device itself and extend to users' ability to shape their experiences of the urban environment. Michael Bull (2008) writes about this extreme form of individualization:

It is a **hyper-post-Fordist culture** in which subjects construct what they imagine to be their own individualized schedules of daily life—their own daily soundtrack of media messages, their own soundspace as they move through shopping centres, their own work-out soundtrack as they modulate the movement of their bodies in the gym (p. 3).

In a **Fordist culture**, by contrast, consumers could design a consumer product with their personal touch through personalized specifications; in a hyper-post-Fordist culture customization goes much further, allowing users to customize their interactions with the urban space. Bull (2008) describes how through the creation of an imaginary sound bubble, the iPod culture allows individuals to regain some control over their urban space. And, as mega-cities continue to grow, the iPod has come to epitomize the possibility of withdrawing from the hectic life of urbanization and retracting in one's own personal and controlled space. The 2001 comedy *Bubble Boy*, directed by Blair Hayes, depicts what life would be like if we lived enclosed in an artificial bubble that separates us from our urban surroundings.

Perhaps more so than other popular contemporary technological devices, such as cellphones, the iPod necessitates acquiring other related technologies in order to be properly used. People must be able, at the very least, to have access to a computer and broadband internet connection. Thus, despite its diverse functional benefits, the iPod is a device whose diffusion, practicality, and adoption is severely limited to societies that are able to readily support a strong, computer-based, and Internet-intensive technological

Do you think MP3 players like the iPod provide people with convenient, custom ways to connect with media and control their interaction with urban space, or do they encourage a withdrawal from urbanization and daily interactions?
Source: © iStockphoto.com/Radu Razvan.

infrastructure, as well as a populace who is able to afford to purchase digital leisure products and to demonstrate an above-average degree of technical sophistication.

Cellphones as Empowerment

The second example of twenty-first-century technology is the cellphone or mobile phone, which lies in stark contrast to the iPod because it is a communication device serving a clear utilitarian purpose rather than being an entertainment tool. The Western mainstream assumption is that technologies have pervaded all aspects of society and are a part of our everyday lives. However, this is clearly not the case for most individuals.

The **International Telecommunications Union (ITU)**, the **information technology (IT)** stream of the United Nations, is dedicated to recording and observing **technological inequality** around the globe. In its most recent study, inequality along the lines of computers and the Internet continues to persist, but has rapidly narrowed when examining cellphone use. Indeed, the penetration rate for cellphones has increased in developing countries from about 10 per cent in 1995 to almost 80 per cent in 2005. This is particularly surprising considering that developing countries continue to lag behind Western countries when it comes to penetration rates of **fixed lines** or land lines (see Figure 2.1). However, the unique characteristics of the cellphone make it a technology that can easily disseminate, even in developing countries, providing a means of empowerment to those with limited economic resources.

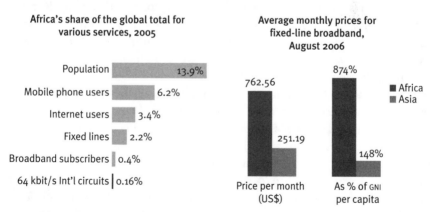

Note: In the right chart, the price sample is based on the 22 African economies that had fixed-line broadband service at the end of 2005. The average value is inflated, since in a high proportion of the economies in this sample, broadband is offered through leased lines and is priced as a business service, rather than for residential users.

Figure 2.1 Use of Communication Technologies in Africa
Source: ITU. 2007. *World Information Society Report 2007: Beyond* WSIS. Retrieved from www.itu.int/osg/spu/publications/worldinformationsociety/2007/WISR07-chapter2.pdf.

Several factors allow cellphone use to spread more widely in comparison to other technologies:

1. *Infrastructure requirements*. Cellphones do not require the same kind of infrastructure that other technologies do, and, most importantly, they do not require wires. Even remote places could potentially get a signal with the availability of a **cellular network**.

2. *Economic factors*. The cost to purchase a cellphone is lower than that to acquire a computer, laptop, or other electronic device, greatly reducing the initial investment in hardware. Moreover, cellphones do not require the payment of monthly fees, like a telephone or an Internet connection does—they can be used flexibly on a pay-per-use basis.

3. *Literacy prerequisites*. Cellphones fit well in oral societies, where literacy levels are low.

4. *Skill levels*. No computing skills are required to use a basic cellphone because a mouse, buttons, and navigation are unnecessary.

5. *Help and support*. In the context of developing countries, technical help for cellphones is easier to obtain than it is for other more complex and less widely available technologies, such as computers, the Internet, and other digital devices.

6. *Cultural norms*. The cellphone fits well with many cultural norms because it allows individuals to maintain their social connections. We will discuss the example of how cellphone use was widely diffused in the Philippines.

Most research examining the diffusion of cellphone use has focused on developed countries, in particular, the United States, Japan, and Norway. Recently, however, with increased penetration of cellphones in the developing world, interest in its impact on these countries has been growing. Unfortunately, the majority of studies produced so far have had a narrow focus, looking at either penetration rates or economic development. The book *Txt-ing Selves* by Pertierra et al. (2003) is unique in that it aims to further our understanding of how Filipinos use cellphones in everyday life, and how the unique cultural context shapes the social effects of this technology on Philippine society.

In 2004, there were 10.5 million cellphone users and 2 million Internet users in the Philippines (Regional Surveys of the World, 2004). In 2009, these numbers increased to 74.5 million cellphone users in comparison to almost 6 million Internet users (International Telecommunication Union, 2009). Pertierra et al. (2003) interviewed ordinary Filipinos to learn about their use of the technology. One case study focuses on Vilma, a retired woman who uses her cellphone as an alternative way to keep in touch with friends and relatives. She feels empowered by the technology because she can now continue to be socially active from the comfort of her home, which reduces her sense of loneliness and increases her self-worth. On the basis of various similar case studies, Pertierra and his team conclude that cellphones provide users with a sense of empowerment and are typically used for social and information purposes, helping to maintain existing relationships, while not used much for creating new ones.[3]

Pertierra et al. (2003) also discuss the role of the cellphone in the context of recent political events. The authors focus primarily on EDSA 2, which refers to the large-scale protest organized at one of Manila's most central avenues, Epifanio de los Santos Avenue. This protest was pivotal for removing former president Joseph Estrada from power, and the media depicted the cellphone as playing a key role in mobilizing large numbers of protesters through text messages. However, Pertierra et al. question the role assigned to the cellphone in staging the protest. They argue that in these discussions the cellphone is perceived as the principal agent of the revolt, ultimately leading to the mobilization of the masses and the demise of ex-president Estrada. In their opinion, a critical analysis needs to go beyond the technology and its capabilities, and ask questions about the origins of *texting*: "who their authors and initial disseminators were, and how such messages facilitated the coordination of the actions" (Pertierra et al., 2003, p. 107). This discussion parallels the debate about the role of social media in the Middle East protests, which we will discuss further in Chapter 8.

Technologies are often characterized in utopian terms as a means to overcome inequalities between rich and poor, educated and uneducated, and to support social movements in their quest for social change. While technologies do provide a means to improve individuals' quality of life, technologies can also increase divisions between the haves and the have-nots, thereby exacerbating the inequality gap, which we will discuss in more detail in Chapter 7. While many extant studies focus on technology's impact on development and economic growth, *Txt-ing Selves* takes a less deterministic view of technology, seeing it as blending in and becoming a natural part of modernity in the Philippines.

Conclusions

This chapter provided an overview of technological achievements from the early stages of humanity to the present. Understanding the close relationship between the introduction of a new technology and social change is not a straightforward task. McLuhan argued that society tends to approach technological developments with a rear-view mirror approach, where "[w]e march backwards into the future" and a new social order is not perceived until it is already in place (McLuhan & Quentin, 2003, p. 68). Hence, we tend to look backward to make sense of how we have changed as a society instead of looking ahead toward how transformations are occurring.

To counteract such a rear-view mirror approach, this chapter presents on-the-ground examples showing how technological transformations intersect with cultural, political, economic, and social change. The examples demonstrate how technologies can be examined as social agents in a sociotechnical system. From this view, technology has transformative power and can shape society. Nonetheless, it is important to keep in mind that technologies alone are not the single causal factor in enabling societal change. We demonstrated how technology shapes society in a cultural, economic, and political context, where many factors come together to create social change. For instance, the divergent effects the printing press had on Europe and Asia demonstrates this coming together of multiple factors to provide the context for social change.

Questions for Critical Thought

1. Can we understand technological developments in the past by looking at them through today's lens? What are the biases that can emerge? How do we take them into account?

2. How does McLuhan's probe "the medium is the message" characterize the effect of media on society?

3. Describe how the iPod feeds into the creation of a hyper-post-Fordist culture and how this is different from a Fordist culture.

4. Think of how you use the cellphone to communicate with friends and relatives. Do you think the technology functions as a means of empowerment? How does it transform your ways of socializing? Do you think this change is best described as revolutionary or evolutionary?

Suggested Readings

Bull, M. (2008). *Sound moves: IPod culture and urban experience.* New York: Routledge. An in-depth investigation into the iPod culture and its ramifications for society.

Haigh, T. (2011). The history of information technology. *Annual Review of Information Science and Technology,* 45, 431–487. This article explores the literature and key ideas in the history of information technology.

McLuhan, M. (1962). *The Gutenberg galaxy: The making of typographic man.* Toronto: University of Toronto Press. A comprehensive analysis of the complex interplay of communication technology, cognition, and society.

Online Resources

Marshall McLuhan, the Man and his Message: CBC Digital Archives
http://archives.cbc.ca/arts_entertainment/media/topics/342/
> The Canadian Broadcasting Corporation (CBC) has put together on its
> website an archive of McLuhan's interviews with the media, including
> radio and television programming.

Maclean's: Mind-bending Mysteries at the Perimeter Institute
www2.macleans.ca/2010/09/17/solving-the-universe/
> An overview of the Perimeter Institute for Theoretical Physics in
> Waterloo, Ontario.

McLuhan Interviews on YouTube
www.youtube.com/watch?v=ImaH51F4HBw&feature=related
> A series of interesting interviews with Marshall McLuhan have been
> posted on YouTube dealing with a wide range of topics related to the
> effect of media on society.

3 Theoretical Perspectives on Technology

Learning Objectives

- ⊛ To address the fundamental difference between utopian and dystopian views of technology.
- ⊛ To compare and critically examine a wide range of theories on the complex interrelationship between society and technology.
- ⊛ To learn about the field of science and technology studies (STS) and its unique socio-technical perspective.

Introduction

A wide range of theoretical perspectives have been proposed that seek to show the ways in which technology and society are linked, as well as the elements of society most affected by technology. The plurality of perspectives shows that there is no single approach for examining the nature of the complex interrelationship between technology and society but, rather, a range of divergent perspectives that each shed light on a different aspect of technological society. A long-standing debate exists between those who see technology as having only positive effects—the utopians—and those who see technology as having primarily negative effects—the dystopians. In addition, early discussions in the field centred on whether technology determines society or society determines technology. In the literature, these two views are referred to as **technological** and **social determinism** because their core assumption, respectively, is that either technology or society has a unidirectional, strong effect that is not, or only minimally, mediated by other factors. We will review the key premises underlying these theories and discuss their strengths and weaknesses.

As these early theoretical perspectives represent a rather simplified view of the technology–society interrelationship, recent perspectives are more complex and attempt to provide a more detailed view of how technology and society intersect. We discuss the field of **science and technology studies (STS)**, which emphasizes that artifacts are socially constructed, mirroring the society that produces them. At the same time, tools shape society as well as its values, norms, and practices (Bijker et al., 1999; Callon & Law, 1997;

Latour, 1993). As a result, social change can only be understood in relation to technological developments. Within the field of STS we review the three most prominent theories: **actor network theory (ANT)**, **social construction of technology (SCOT)**, and **social informatics (SI)**.

Utopian versus Dystopian Views of Technology

One of the main perspectives on technology at the most basic level can be categorized as the **utopian perspective**. Utopians embrace technology as a new means of achieving progress and efficiency. The idea of progress emerged in the seventeenth and eighteenth centuries, particularly in Europe and North America, and became something desirable, a sign of civilization. The utopian view reflects the assumption of technology as "the realization of science, revealing itself in ever increasing control over nature" (Street, 1992, p. 20). In this view, technology allows us to dominate and manage nature, making our lives easier, and it leads to advancements in how we produce material goods, thereby reducing costs and increasing efficiency. Progress is not limited to the production of material goods but also to the accomplishment of societal goals, including higher levels of security, better means of communication over time and space, improved health care, and increased autonomy (Hill, 1989). Overall, "technical change serves to improve the quality of life" (Street, 1992, p. 20) and makes many aspects of life easier.

The other main perspective of technology is the **dystopian perspective**. For Street, technology "threatens established ways of life" and is thus seen as a **regressive force** (1992, p. 20). This line of thought was already present in early Industrial Revolution England, where a group referred to as the **Luddites** formed to counter changes occurring in work conditions as a result of mechanization (see Chapter 6). Based on this rebellious movement against technology, the term *Luddite* or *neo-Luddite* refers to individuals who resist technological developments.

While society has tended to embrace technology as a means to improve life, "[f]or the generation of the interwar years, blind faith in technology was not possible" (Francis, 2009, p. 14). This was primarily because this generation experienced the negative consequences technology can have when applied toward destruction, extermination, and death. World War I and World War II showed how technological developments had radically changed combat, introducing new dimensions to conflict, devastation, and human suffering.

While not all scholars advocate for simply embracing dystopian views of technology, many scholars do argue that the study of technology needs to go beyond a focus on the immediate and intended benefits of technological inventions, and should also analyze the often **unintended effects** that result in the long run, for example, the effects of industrialization and globalization on the environment.

Tenner (1996) sees many technological solutions to societal problems as creating new problems, often larger in scale. He has termed the unintended consequences resulting from technology as **revenge effect** and through a series of case studies analyzes these problems in depth. An example of a revenge effect is when a drug to induce hair growth, under certain conditions, accelerates hair loss. If the likelihood of cancer had been raised by taking the drug, this would be considered a **side effect** rather than a revenge effect. Side effects occur in an area unrelated to the one the technology was supposed to function within, whereas revenge effects occur directly in the area of intervention but are unintended.

The utopian and dystopian viewpoints provide a rudimentary understanding of how technology affects society. In the next section, we present the various theories that examine in more detail how technology and society intersect.

Theories of Technology and Society

Numerous theories of technology and society have been proposed. To examine these varied theories in more detail, Feenberg (1999) has developed a theoretical model (see Figure 3.1), in which he distinguishes between two central dimensions: (1) neutral (as in the **neutrality of technology argument**) versus **value-laden**; and (2) **autonomous** versus **human controlled**. In this section, we first describe the two dimensions and then discuss in depth each of the four theoretical frameworks that result from the two-by-two matrix depicted in Figure 3.1.

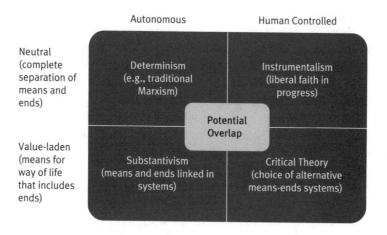

FIGURE 3.1 Theories of Technology and Society
Source: Adapted from Feenberg, A. (1999). *Questioning technology (p. 9)*. New York: Routledge.

Neutral versus Value-Laden

The first dimension encompasses theories that view technology as falling in the category of either neutral or value-laden. Theories of neutrality describe technology as separate from human activity and with no effect on natural ends—that is, on the fundamental elements of human nature (e.g., ethics, morality, forgiveness, and happiness). Feenberg argues that this neutralization of technology hinders any in-depth analysis of social change because "if technology merely fulfills nature's mandate, then the value it realizes must be generic in scope" (1999, p. 2). An example of the neutrality of technology is the debate around the value of guns, which is discussed in Box 3.1.

Box 3.1 gives an example of the arguments put forward by groups supporting the view of technology as neutral. The opposing view—that technology is value-laden—tends to equate technological development with human progress. In this view, any new experience realized through technology is seen as progress for the entire human race. Feenberg (1999) shows how often the first landing on the moon is described as a collective achievement: *we* as humans landed on the moon. This removes any sociological, cultural, and economic analysis from the technological experience, and buries the political consequences that lie behind many technological developments. For example, there is no mention in the retelling of the first landing on the moon about the arm's race taking place between the United States and the USSR, or about the funding that supported, and continues to support, space travel and exploration. This is taxpayers' money that is used with the aim of advancing science and technology, at the expense of education or health care.

Autonomous versus Human Controlled

The second dimension of Feenberg's model distinguishes between technology as exerting control and technology as controlled by humans. The concept of technical development is at the centre of this discourse. Technological autonomy is based on the idea that technological developments and inventions are guided by independent or self-serving dynamics, separate and distinguishable from societal or human influences. Although human beings are involved in the creation of technologies, supporters of **autonomous technology theories** would argue that humans have little choice in deciding how the technology will evolve and diffuse in society. The underlying assumption of this theory is the notion that technology propels and alters the development of social structures and cultural values.

Within this view, once technologies are introduced into an environment, they will be the key factor in determining the direction of social change and progress. Once fixed within a society, the corresponding roles and actions associated with the technology become more normalized, thus limiting the input of **human agency**. For instance, the introduction of television is

Box 3.1 ❖ The Neutrality of Guns

In the neutrality view, technology is neither good nor bad in and of itself; instead, it is the means by which humans employ technology that gives it meaning. An often cited example is the way that guns impact society. Many groups, such as the National Rifle Association of America (NRA), argue in favour of individuals having the right to own a gun because in their view guns are neutral and have no consequences or moral value. Guns are merely machinery; it is the individuals who use them who can act in good or bad ways.

With a dramatic increase in deaths in the Mexican city of Ciudad Juárez, the Obama administration has been advocating in favour of introducing further laws to limit the availability of guns to gangs operating in the area. The NRA's website provides a short summary of the organization's position toward any kind of interference linked to gun ownership by arguing that it is not the guns that are creating the problem but, rather, the political situation in Mexico itself and that any kind of further regulation would only be an infringement on Americans' individual rights.

How congruent or incongruent is the imagery used on the cover of this NRA digital publication with the organization's message that guns are simply pieces of machinery with no consequences or moral value ?
Source: © NRA Life of Duty Network

The following is a message from Wayne LaPierre, the NRA executive vice-president, regarding the organization's point of view on establishing new regulations around gun ownership:

What They Didn't Tell You: Obama's Registration Scheme

For months now, Mexican President Felipe Calderón has used our gun laws as a scapegoat for the violence caused by the drug cartels in his country. Once again, it seems our own president is joining in the chorus of blame. Early in the Obama administration, Attorney General Eric Holder called for a new ban on semi-automatic rifles, only to be rebuffed by 65 members of his own party. Knowing that he can't push an anti-gun agenda through Congress, Obama now wants to use the power of the executive branch to issue a new "emergency requirement" for thousands of federally licensed firearms dealers.

This so-called emergency has been going on in Mexico for four years, and has nothing to do with our gun laws. El Paso, Texas, is the safest city in the nation. Just across the Rio Grande, in Ciudad Juárez, 3,000 people have been killed this year. This is just a shallow excuse to engage in a sweeping firearms registration scheme, and the NRA will do everything in its power to stop these "emergency requirements" from taking effect. (http://home.nra.org/#/home, 22 December 2010)

perceived as a catalyst in re-configuring leisure, communication, consumer, and cultural practices in the twentieth century. Supporters of this theory would argue that television as a technological medium became a focal point in family and social life. It changed the manner in which information was acquired and processed, and also created mass communal events through which people have indirectly and simultaneously shared analogous experiences. Programmed events such as the Super Bowl or the series finale of a popular television show, such as *M*A*S*H** or *Seinfeld*, could be considered examples of this phenomenon.

By contrast, critics of the autonomous viewpoint argue that technology is a socially constructed entity, whose meaning and use is determined by human action. Rather than being diminished by technology, human agency becomes the central ingredient in understanding and determining the role of technology within its social context. Human beings have a choice in selecting and deciding how technologies will be used, as well as determining the value given a particular technology. Concepts showcasing this viewpoint, such as the social construction of technology (SCOT, discussed below), espouse the belief that technology is shaped by human needs and social factors. The cultural norms and values within a social system influence the construction, diffusion, and utilization of the technological product (see Chapter 5). This reasoning can be used to suggest why certain technologies fail to diffuse in cultures or societies that are indifferent toward adopting a product or system that is incompatible with their social beliefs or structures.

Four Theories of the Technology–Society Intersection

As indicated on page 43, Figure 3.1 shows the four main theories identified by Feenberg (1999) along the two dimensions of interest previously discussed, forming four quadrants, each representing a different vision of technology's intersection with society. Each will be discussed next and on-the-ground examples provided.

1. Determinism

Determinism is divided into two primary and opposing theoretical views: technological and social determinism. **Technological determinism** proposes that technology is the driving force in developing the structure of society and culture (see Figure 3.2). Technological determinists adhere to the notion that technology directs and shapes social interactions and systems of thought. The uses of technology are dictated by the design of the technology itself. Technological determinists view technology as an independent and autonomous entity guided by its own internal logic. In response to technological developments, society changes its institutions, modes of communication, labour practices, and cultural meanings. Moore's Law is an example of the view of technological determinism. The law stipulates that "the number of transistors that could be placed on an integrated circuit would continue to double at short, regular intervals has held true ever since, although the interval soon stretched from twelve to eighteen months" (Ceruzzi, 2005, p. 584). As the speed and capacity of computers increases, humans need to adapt to and keep up with these changes. And since technology is self-directing, it is not influenced by social or cultural factors within society.

The second perspective—social determinism—sees factors in society as creating specific uses of technology (see Figure 3.1). Social norms, attitudes, cultural practices, or religious beliefs would directly impact how technology is used and what its social consequences are. For example, research into the use of media for communicating during a breakup found that norms around communication impact young people's perceptions of media adequacy. Text messaging and social media, for instance, were not perceived as adequate for transferring complex messages during a breakup, whereas in-person communication was perceived as more adequate (Gershon, 2010). We will discuss the impact of social norms on communication during breakups in more detail in Chapter 9. This example illustrates the social determinism view, which argues that societal or group norms will determine how a technology

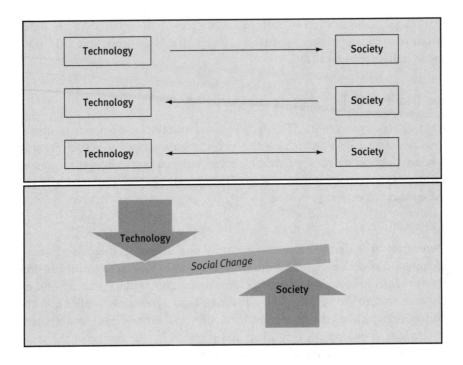

FIGURE 3.2 Technology and Social Change

is utilized. To summarize, technological determinism views technology as the driving force of social change, while social determinism views social factors as affecting technological development and use.

2. Instrumentalism

Instrumentalism analyzes technology as a neutral tool or instrument, whose purpose is to fulfill users' specific tasks. Instrumentalists believe that technology can be understood as an evolutionary process in which technologies are the product of previous technological endeavours, whose rationale is to improve productivity and efficiency. Since technology is characterized as merely being a neutral tool, **instrumentalism** proposes that technologies can be used for either positive or negative reasons depending on the moral intentions of the human agents who employ them. This is exemplified in the perspective the NRA takes on guns, discussed in Box 3.1.

3. Substantivism

Substantivism argues that technology brings forth new social, political, and cultural systems, which it then structures and controls. This view supposes that technologies operate under their own inherent logic and goals. Within this concept, the essence of technology is both externalized and unable to be controlled by human beings. A substantivist position expresses the view that technology can act as an independent force. Acting as a neutral character, technology shapes and dictates society. Propelled and guided by its own embedded logic and goals, technology has the ability to modify cultural conditions. For instance, in Chapter 2 we discussed the mass media and its effect on cultural production. The embedded logic of the mass media is to attract large audiences and to commodify cultural goods. While a circus is a local performance that attracts a pre-established group of people, cinema can reach a mass audience independent of space and time. This example demonstrates how technologies are not neutral but, rather, encompass specific social and political values.

In contrast to instrumentalism, substantivism assumes that technologies can be used for either liberating or destructive means, according to the nature of the technology itself, which establishes and controls society, rather than according to the means and goals of human actors. Human agency, then, cannot control technologies; instead, the technology's inherent logic and goals determine its use. From this perspective, the logic and goal of a weapon is to harm and perhaps even to kill. By contrast, the logic and goal of the printing press is to diffuse information. These examples show how the nature of technology predetermines how the technology will be utilized and the kinds of impact it will have on society. Substantivism gained social relevance during the Cold War, when the atomic bomb made the possibility of mass destruction very real. Atomic weapons shaped society, increasing fear and impacting people's perceptions of safety.

4. Critical Theory

Critical theory suggests that technology is the product of both technical and social factors. Technology is not simply a means of satisfying goals, but a process that directs a specific mode of living and understanding. Technology is not viewed as a symbol of linear progress but, rather, as an element capable of adopting different possibilities and directions depending on the social influences and values of its users. Technology must be understood, then, within the context of its use and development. When governed by a **technocracy**, technology embodies the values, social structures, and goals of **hegemonic elites**. That is, the ruling class dominates through the use of technologies that subserve their needs. By contrast, participatory democratic actions offer an alternative to technocracy through "tactical resistance to established designs [that] can impose new values on technical institutions and create a new type of modern society" (Feenberg, 1999, p. 128). Through resistance, technology can be used toward democratic aims, instead of technocratic interests reflecting the values of the social system that makes use of the tool. Chapter 10 discusses the example of counter-surveillance to illustrate the use of technology to question existing forms of domination in society and to establish new norms.

Science and Technology Studies (STS)

Science and technology studies (STS) is an interdisciplinary field concerned with the study of how scientific and technological changes intersect with society. Early STS theorists such as Ellul and Vanderburg (1964; 1981) and Mumford (1967; 1981) argued that technology was an independent force dominating society and thus requiring some form of control and delimitation (Cutcliffe & Mitcham, 2001). STS, however, has abandoned the view that technology is an autonomous force at work independently from society. The aim of STS is to analyze technology in its unique social, local context of "complex societal influences and social constructs entailing a host of political, ethical, and general theoretical questions" (Cutcliffe & Mitcham, 2001, p. 3). This fits perfectly with the definition of technology proposed in Chapter 1. Technology in this context is conceptualized as value-laden, actively shaping and in turn being shaped by culture, politics, and social values (Cutcliffe & Mitcham, 2001). Table 3.1 provides a summary of key characteristics of STS approaches (Bijker et al., 1999).

STS confronts two major challenges (Cutcliffe & Mitcham, 2001). First, STS aims at conducting meaningful research both at the micro and macro levels. Hence, STS needs to combine individual and global levels of analysis to be able to have real impact on society's understanding of technology as well as policy. The second major challenge is to neither embrace a utopian (promotional enthusiasm) nor dystopian (anti-science and technology) view but, rather, to provide in-depth analyses of technological design, implementation, use, and social consequences. We discuss three approaches within the field of STS that are representative of the field: social construction of technology (SCOT), actor network theory (ANTS), and social informatics (SI).

TABLE 3.1 **Key Characteristics of STS Approaches**

Characteristic	Description
1. Rejection of technological determinism	STS rejects the notion of technological determinism, where technology is perceived as the agent of social, cultural, political, and economic change.
2. Rejection of social determinism	STS also rejects the notion of social determinism, where the inventor of a technology alone drives technological progress without any consideration of the social system in which an invention occurs.
3. Holistic approach	STS intends to study the entire socio-technical system and not the social, political, cultural, and economic dimensions separately. That is, social factors are examined in relation to use or non-use of technology.
4. Qualitative methods	STS uses qualitative methods, such as case studies and ethnographies, to provide an in-depth examination of socio-technical systems that generate rich descriptions.

Source: Based on Bijker et al., 1999.

Social Construction of Technology (SCOT)

Social construction of technology (SCOT) emerged in the 1970s and its roots are evident in the sociological theories influencing academia during this period. In particular, the humanities eschewed a top-down view of history, politics, and the arts in favour of a bottom-up approach that sought to understand the social constructs and relationships underpinning everyday life.

SCOT advocates, such as Wiebe Bijker and Trevor Pinch, do not see technology as shaping human action but instead see human action as shaping technology. Social constructivists argue that a technological object can acquire different uses and values according to the social context it is placed in. For example, the 1980s movie *The Gods Must Be Crazy* provides a compelling depiction of how an object that is so normalized in our Western society and has a particular pre-defined function can suddenly take on multiple meanings when placed in a different social and cultural context. Here a Coca-Cola bottle is dropped from the sky by a passing airplane and makes its way into a Bushmen community in Africa that has had little to no contact with Western civilization. For the Bushmen, the Coca-Cola bottle becomes an artifact with an infinite number of meanings (because they had not encountered one before), including practical (e.g., to fetch water, to hammer, and to play games), symbolic, and religious.

Scholars employ SCOT to understand technical change, the design of tools, and the technology–society relationship. Four key terms help understand the interplay between design, technology, and society. Within SCOT four key concepts emerge for analyzing technology: (1) the **relevant social group,** (2) **interpretive flexibility**, (3) **closure** and **stabilization**, and (4) **wider context**.

1. Relevant Social Groups

Relevant social groups are important due to their influence in attributing meaning to an artifact (Bijker, 2009). Meaning is obtained through interacting with like-minded social groups, who share a similar opinion about the artifact and its uses. Pinch (2009) has argued that "[i]t is the meanings given to technics and techniques that provide a way of understanding successful, failed, tangential, and niche-market technologies" (p. 46). Without the necessary societal support, a new or existing technology can fail to be viewed as useful within a respective group, causing both new and older products to be viewed as obsolete. An example of this can be seen in the ever-changing world of software. For instance, an earlier system, such as Windows 95, no longer holds value in contemporary Western society because it lacks the level of functionality required to support the type of socially necessary services currently available. Modern Internet browsers, word-processing suites, and high-definition videos are examples of technologies that have produced several generations of products, with newer versions rendering older ones obsolete.

2. Interpretive Flexibility

Interpretive flexibility describes how artifacts are not neutral; instead, their meaning emerges in a socio-cultural context. For Pinch and Bijker what this means is "not only that there is flexibility in how people think of or interpret artifacts but also that there is flexibility in how artifacts are designed" (1987, p. 40). Our earlier example of the movie *The Gods Must Be Crazy* also provides an example of interpretative flexibility. The Coca-Cola bottle creates major upheaval in the community as the bottle is repurposed for a variety of tasks, including curing snakeskin, making music, and as a mortar for food preparation. Because community members feel that this bottle has become central to their work, and there is only one bottle, it leads to conflict and controversy. The movie makes it clear that a simple tool, such as a bottle, can be used for multiple purposes and its meaning and relevance emerges in a socio-cultural context.

3. Closure and Stabilization

As an artifact gains prominence in society, its flexibility to be interpreted for other uses decreases because the social construction of the artifact's meaning becomes solidified. *Closure* describes the moment in the cycle of design when the relevant social group has reached a consensus of what the tool is all about. When stabilization is reached, the tool has been assigned a very specific use and little experimentation occurs as to what other purposes the tool could serve.

Pinch and Bijker further distinguish between **rhetorical closure** and **closure by redefinition of the problem**. Rhetorical closure occurs when a technological problem is not solved but a solution is presented by changing the language used to describe the problem and hence modify its social meaning. For instance, reports in the media have suggested that certain models of Toyota vehicles are unsafe because of technical problems they had been experiencing.[1] While these problems could require a technical solution, the problem could also be solved by convincing users that the cars do not represent a danger because they are technologically sound. Here the solution to the problem does not require a technical change to the artifact itself but, rather, new rhetoric about safety. In closure by redefinition of the problem, again a technical problem is not solved, but instead the problem is reformulated in different terms. In the Toyota example, this would mean that instead of solving any mechanical or electronic problems, the problem would be linked to how people interact with and respond to the car and not the functioning of the car itself. This shows that problems linked to technology often reflect how people interact with and make sense of these technologies.

4. Wider Context

The last term is *wider context,* which describes how "the sociocultural and political situation of a social group shapes its norms and values, which in turn influence the meaning given to an artifact" (Pinch & Bijker, 1987, p. 428). Norms and values are powerful frameworks for interpreting artifacts and for understanding their value in society. A prime example is birth control. Even though birth control is a safe and effective method with widespread popularity in many parts of the world, it is rejected in some regions because it is in opposition to the culture's values and norms. This clashing of values and technology will be discussed in the next chapter under the umbrella term *compatibility.*

Pinch and Bijker's (1987) primary example for demonstrating the social construction of technology is that of the bicycle. Early bicycles such as the high-wheeled ordinary or penny-farthing bicycle were easily recognizable due to their prominent and large front wheel and small rear wheel. These bicycles were used primarily by young men for the purposes of sport, whereas women were openly discouraged from using these bicycles. However, these early models were racked by safety issues, especially vibration and braking. These factors led to a redesign of the bicycle that served the interests of various groups of people (young men, women, and the elderly). The introduction of smaller air tires improved safety for female and elderly cyclists. Engineers viewed the air tire as a "theoretical and practical monstrosity," whereas users over time saw its practicality and value by simultaneously boosting speed and safety (Pinch & Bijker, 1987, p. 427). Box 3.2 uses the case study of the electric car in France to further exemplify some of these points.

Box 3.2 demonstrates how the development of the electric car can be examined using SCOT. The strength of this approach is that it provides a useful framework for understanding how design unfolds over time and helps identify key social factors.

Nevertheless, critics of SCOT, such as Langdon Winner, have argued that the theory lacks an understanding of "the dynamics of technological change," such as the social context, economic conditions, and structural relationships within society at the time of a particular technology's introduction (Winner, 2003, p. 234). Second, critics have argued that SCOT supporters spend too much time studying the development and social construction of a technology, while at the same time showing "a disregard for the social consequences of technical choice." Winner (2003) characterizes these as the **social after-effects** that result from selecting or adopting one technology over another, such as the transformation of "personal experience and social relations" (p. 237).

BOX 3.2 ❖ THE CASE OF THE ELECTRIC CAR IN 1970S FRANCE

Despite being prophesized for several decades, the age of the electric car has yet to arrive. With its low fuel costs, quiet hum, and clean emissions, theorists have long proposed the advent of the electric car to be imminent. Yet while cars powered by hybrid fuel cells begin to make their mark on the North American automotive landscape, the emergence of the electric car continues to be more science fiction than fact. As discussed by Callon (1987), the development of the electric car is an interesting example of how the success of an emerging technology can be heavily weighted by the social factors and shifting values constructing its meaning.

Callon's (1987) work analyzed the attempt of several parties, who bridged the worlds of science, industry, and government, to combine their knowledge to try to create an electric car in 1970s France. The plan was devised by the French state electricity company Electricité de France (EDF) as a transitional stepping stone from an industrial to a post-industrial age. In response to social protests and spiralling fuel costs, EDF envisaged "a society of urban post-industrial consumers" who shared the notion of ushering in a "new era of public transport" driven by the cost-effective and environmentally friendly electric car (Callon, 1987, p. 85). Understanding the status that private car ownership symbolized, EDF hoped that science and technology would be put at "the service of the user and to do away with social categories that attempted to distinguish themselves by their styles of consumption" (Callon, 1987, p. 85).

To manufacture the vehicle, the Compagnie Générale d'Electricité (CGE) would design the electrical components, such as the motor and batteries, and car manufacturer Renault would assemble the vehicle's body and structure. Problems began to arise within the process when the cheap catalysts envisioned to run the vehicle became contaminated. Furthermore, Renault started to actively critique the project.

The issues identified by EDF as signs of an imminent post-industrial rejection of the traditional car, such as poor mileage and pollution, were viewed instead by Renault as displaying temporary antipathy "with the car industry's lack of dynamism and the state of public transport" (p. 91). Renault's solution involved using better public buses, which provided passengers with greater comfort and less harmful emissions. Furthermore, in response to escalating fuel costs, the general public's view of the traditional car had been "rehabilitated" due to the car's better fuel performance and fewer emissions (Callon, 1987, p. 91). As unemployment rose and social protests declined, interest in the electric car rapidly diminished. Without support from relevant social groups, interest in the project died and the traditional car in its new guise took on a different meaning.

A third area of criticism is the way in which SCOT scholars give importance and salience to the opinions of some groups and interests over others. That is, SCOT overlooks those groups within society that have no input in approving a technology or that suffer from the social consequences of that technology's selection (Winner, 2003, p. 238). By disregarding these groups, SCOT only provides a limited perspective of how technologies are diffused and how they gain social relevance in society. Additionally, the relativist nature of the SCOT approach—in which each social group gives meaning to a technology—provides little information for determining whether one social construction is better or more valid than another (Boreham, Parker, Thompson, & Hall, 2008).

Actor Network Theory (ANT)

Actor network theory (ANT) is a sociological theory popularized in the 1980s by scholars Latour, Callon, and Law. This theory treats "everything in the world as a continuously generated effect of the webs of relations within which they are located" (Law, 2009, p. 141). Proponents of this approach examine and describe the relationships and practices undertaken by actors. Within ANT, **actors** or **actants** can emerge in a variety of forms ranging from human beings to concepts or ideas, technologies, institutions, and so forth. Rather than explain the reasoning as to why relationships between actors occur, the theory instead examines how these relationships are constructed and practised—that is, ANT is based on a constructivist approach, where relationships are in a particular social context. A strength of this theory is its ability to examine the active processes and interconnected relationships between human and non-human actors, which is not possible in many of the other theories of technology.

At the core of understanding these relationships is the concept of the **network**. Networks consist of relationships between people, but they also comprise interactions with objects and organizations. In Latour's classification of the term *network*, it contains "resources that are concentrated in a few places" and that are connected to one another (1987, p. 180). Subsequently, the "connections transform these scattered resources into a net that may seem to extend everywhere" (Latour, 1987, p. 180).

In contrast to the contemporary usage of the word to characterize the transportation of information, ANT views the role of the network as one of **transformation** (Latour, 1999). Networks are categorized by their size, concentration of resources, and fragility. Using as an example telephone lines, Latour notes that individual lines are "minute and fragile" to the point that none are rendered on a map and each line can be easily cut (Latour, 1987, p. 180). However, as a group, the telephone lines seemingly encompass a vast network, crossing geographical boundaries and functioning as a single unit. Nevertheless, if a single element of the network is fractured or disabled, the network is prone to failure. To remain strong, networks undergo a continuous process of re-evaluation and redevelopment, which includes being influenced by elements from other networks. For instance, the Internet has changed the telephone network in fundamental ways.

In the realm of technology, ANT aims to "transcend the distinction between the social and the technical" through an impartial approach (Grint & Woolgar, 1997, p. 30). Relationships within ANT are heterogeneous in their composition, and, as mentioned, include both human and non-human participants. These associations enable change to take place through collaborations between science and technology, which is possible due to the knowledge held by those participating within the network, such as scientists and engineers. Within laboratories, human actors can test the success of alliances between particular items and tools. However, the success of a technical breakthrough is not merely due to either the human actor or the tools but through the manner in which their collaborative relationship is constructed.

Latour (1988) suggests that the translation of the laboratory to the unfamiliar environment of the field was crucial to Louis Pasteur's success. Equipped with laboratory equipment, such as microscopes and logbooks, the Pasteurians engaged in fieldwork, learning from people such as farmers and veterinarians, in order to be able to determine which conditions and factors were leading toward the disease and potential causalities. The results of Pasteur's trials "were to create a new object that retranslated the disease into the language of the laboratory" (Latour, 1988, p. 76). Box 3.3 shows how Latour employed ANT to uncover the complexity of the process that led to Pasteur's discoveries.

When viewing Pasteur's triumphs through the prism of ANT, responsibility for his discoveries shifts from a single individual to the multitude of actors, human and non-human, whose respective contributions impacted the development of scientific discoveries. Thus, when studying the history of technology from a sociological perspective, we can use ANT to understand and acknowledge the roles, systems, and networks necessary in the development and promotion of new technologies.

Despite its merits, however, scholars have identified several weaknesses. First, a major point of criticism for the opponents and detractors of ANT is the heterogeneous nature of its actors and networks. By positing human and non-human actors on the same footing, scholars have criticized ANT for putting too much emphasis on the role of non-human actors, thereby reducing human actors to mere objects. This is particularly relevant when detailing the role of power in shaping, re-casting, and ending alliances between human and non-human participants (Grint & Woolgar, 1997). The role of power in defining the relation between human and non-human participants was already discussed in Chapter 2 when we

BOX 3.3 ❖ DISCOVERING PASTEURIZATION

French chemist and microbiologist Louis Pasteur is renowned as being one of the most important scientific minds of the nineteenth century. Innovations in microbiology attributed to Pasteur include vaccines for anthrax and rabies, the necessity of sterilizing medical equipment, as well as the process that bears his name: **pasteurization**. The latter, arguably Pasteur's most celebrated achievement, assisted with the elimination of pathogens found in various foods and beverages, such as milk and cheese, which can cause disease.

Responsibility for the success of these scientific and technical breakthroughs has long been solely accredited to Pasteur. In the concept known as the **Great Man Theory**, history is often written and mythologized in overly simplistic terms, where great events are attributed to remarkable individuals, often men, who demonstrated intellectual or political prowess during their lifetime. Yet such a viewpoint neglects the myriad socio-economic, political, cultural, or technical transformations that may have influenced the direction and choices that these individuals made. Additionally, the focus on a single person ignores the input of assistants and helpful associates, and the transmission of knowledge and information from external sources necessary to the work of brilliant thinkers such as Pasteur.

Commenting on feats linked to Pasteur, Bruno Latour (1988) has argued that "[i]f the whole of Europe transformed conditions of its existence at the end of the last century, we should not attribute the efficacy of this extraordinary leap forward to the genius of a single man" (p. 15). Using ANT as a guide, one instead can view Pasteur as an actor whose technical accomplishments were the product of a network of laboratories, logbooks, and scientific associates. Enrolling the services of likeminded individuals within the scientific community, Pasteur was equipped to translate the interests of other actors in his network, including farmers. The members of Pasteur's laboratory were then able to address the multitude of ideas and issues held by the various actors and render these concerns into a language that was usable within the setting of the laboratory in order to combat the non-human actor, that is, the microbe.

examined how the nature of work, and the relation between craftspeople and their craft, drastically changed as a result of the introduction of new weaving and spinning machines in the beginnings of the industrial era. In this instance, weaving machines did not remove craftspeople from their jobs, but employers used their position of power to replace highly skilled craftspeople with a larger, inexpensive, and unregulated workforce (Berg, 1994). Second, critics have also suggested that ANT fails to properly define or describe the nature of its networks, which can appear overly complex and abstract. Third, critics have opined that the theory tends to acknowledge localized networks and fails to examine either the role of broader macro-level social structures or the influence of cultural practices in affecting the construction and redevelopment of a network (Walsham, 1997).

Social Informatics (SI)

Social informatics (SI) is based on a social constructivist view of technology, which sees technology as emerging in close interaction with society. Social informatics is a "multidisciplinary research field that examines the design, uses, and implications of information and communication technologies (ICTs) in ways that account for their interactions with institutional and cultural contexts" (Kling, Rosenbaum, & Hert, 1998, p. 1047). The broad definition of society as the "institutional and cultural contexts" has allowed for a multitude of approaches and research questions to fall into this category. In this view, technology does not create uses in society nor does society create uses in technology; instead, they mutually influence one another.

The key ideas of social informatics are as follows:[2]
- to conduct a contextual and empirical analysis of the design, implementation, and uses of ICTs;

- in studies, to include social variables in the analyses, such as cultural, organizational, and contextual variables;

- to exam technologies in the context of work processes and practices;

- in studies, to focus on specific ICTs as they are being designed, adopted, and used in particular settings (studies often underestimate the costs and complexities of implementing ICTs and "overestimate the generalizability of applications from one setting or team of individuals to another") (Kling, 1999, p. 228); and

- to view ICTs as not just tools but as objects situated in social systems that thus interact with actors, settings, and norms.

Social informatics aligns with the key premises put forth in the area of STS, which argue that ICTs should not be conceptualized as external forces that impact users and their environments but, rather, as an integral part of users' everyday lives (Wellman & Haythornthwaite, 2002). This viewpoint investigates the use of technology from the everyday experiences that users have with the technology. Studies that see ICTs as incorporated into everyday life examine these technologies in the context of other media and not as a separate sphere. Furthermore, as many technologies become incorporated into everyday practices and social environments, they come to represent the norm rather than an external force.

Conclusions

Simple approaches to understanding the interrelationship between technology and society have been largely refuted. Utopian and dystopian visions of technology are too simplistic and do not consider how technology has become embedded in our everyday practices. Technological determinism views the effect of technology on society as unidirectional and does not consider social factors. By contrast, social determinism assumes that social factors drive technological design, use, and consequences alone without ascribing technology any power. Not even approaches advocating soft determinism provide a satisfactory answer as to how technology and society intersect. Current understandings of technological society propose a mutual shaping process, where technological factors impact society and in turn societal factors impact technological design, development, implementation, use, and social consequences. In these approaches, technology is not studied as a universal force; instead, technologies are examined in unique social contexts with specific cultural, religious, and political characteristics.

Another important factor in understanding the interrelationship between technology and society is globalization. Social change occurs in a global context, where technologies diffuse across borders, cultures, and social groups. As technologies become more ubiquitous and integrated into our daily social practices, routines, and ways of being, continued critical, in-depth, and contextualized analysis of our technological society will be necessary.

Questions for Critical Thought

1. When comparing utopian and dystopian views of technology, which one do you think has more applicability to society today? Provide a rationale for your point of view.

2. What are the main criticisms put forward against the actor network theory (ANT) approach?

3. Pinch and Bijker show how the social construction of technology (SCOT) approach can be utilized to examine the development of the bicycle. What other examples from your own life can show the usefulness of the approach for examining modern technological developments?

4. Can certain theoretical concepts, such as social constructivism or technological determinism, explain the scope of technology in society on a universal level? Or are they better suited toward understanding particular types of societies?

Suggested Readings ....

Feenberg, A. (1999). *Questioning technology.* New York: Routledge. The book provides a critical analysis of the heavy reliance of society on technology and outlines key theoretical perspectives on the complex interlink of society and technology.

Mitcham, C. (1994). *Thinking through technology: The path between engineering and philosophy.* Chicago, IL: University of Chicago Press. The book provides a critical and comprehensive understanding of the philosophy of technology movement.

Pinch, T., & Bijker, W.E. The social construction of facts and artifacts: Or how the sociology of science and the sociology of technology might benefit each other. *Social Studies of Science, 14,* 399–441. The paper describes an integrated social constructivist approach toward the study of science and technology. From this perspective, science and artifacts are defined as social constructs.

Tenner, E. (1996). *Why things bite back: Technology and the revenge of unintended consequences.* New York: Knopf. The book stresses the potential of technology for unintended consequences, without falling into dystopian rhetoric and completely rejecting technology.

Online Resources ....

Society for Social Studies of Science (4S)
http://4sonline.org/
> This is an international non-profit organization dedicated to the study of science, technology, and medicine, with an emphasis on how these develop and interact with their social contexts.

Computer History Museum
www.computerhistory.org/
> The world's most comprehensive institution dedicated to the history of computing and its impact on society.

TED (Technology, Entertainment and Design) Talks
www.ted.com/talks
> This website consists of live streaming and archiving of talks by leading scholars, practitioners, and technologists that take place around the world on a wide range of topics on technology, science, culture, and research.

4 Techno-Social Designing

Learning Objectives

- To look at technological design and how it intersects with society.
- To trace the development of technopoles and uncover their social significance in the modern city and economy.
- To investigate the concept of research and development (R&D) and the pressures existent in the sector.
- To examine the inner workings and outer pressures of software development teams.

Introduction

Users of technology are, for the most part, oblivious of the stages that precede adoption and use. Most users do not consider the creative processes that underlie technological invention and innovation because these processes do not directly affect their day-to-day use of tools. The aim of this chapter is to uncover these often hidden creative processes and to examine the visions of developers, the challenges experienced in research and development (R&D), and the complex interweaving of technological development and societal factors. What factors led to the design of Facebook? How did Mark Zuckerberg make choices about which features to include and exclude in the social networking site?

In addition to examining R&D, in this chapter we introduce the term *technopole* to describe specialized cities dedicated solely to technological innovation. Silicon Valley is presented as an instance of a technopole that combines a highly educated workforce with military and economic interests. We discuss the political, social, and economic repercussions of technopoles as well as their vulnerability in the context of fragile global markets.

The chapter continues by explaining the concept of R&D, the pressures existent in R&D teams, and the ways that innovation occurs in these teams. We then directly link R&D to economic development and review Schumpeter's classic economic model, which argues that the structure of society is directly linked to creative processes and innovation. Next, we present data that compare expenditures on R&D across nations and examine how these are linked to global economic development. The final part of the chapter examines in more detail the inner workings and the outer pressures of software development, which is one type of R&D that has come to occupy a central role in the world economy. As part of the discussion on software development, we examine how technological design can encompass social features. For this purpose, we explain the terms *object affordance* and *social affordance* in order to better understand the complex interplay of technological design and social processes.

Technological Design and How It Intersects with Society

The study of technology tends to focus on technologies that are widely used in society without giving much consideration to the **creative processes** that take place in the design and implementation phases. However, technological design occurs long before a technology is revealed to users. Extensive testing, developing, and prototyping take place during the design phase of a new product, which, because of the high complexity of this phase, often extends over a long period of time. For example, the first developments of the cellphone were described as early as 1945, when J.K. Jett, the head of the US Federal Communications Commission (FCC), announced that AT&T had for the first time used a type of low-powered transmitter that employed high-band radio frequencies that could allow millions of users to communicate (Farley, 2005). However, cellphones did not become widely used in the United States until the 1990s and did not really take off until around 2002.

What are the reasons for the general public's lack of knowledge about the field of technology design? Why do scholars not have a better understanding of how design unfolds if it often covers such a lengthy period of time? Several important factors are responsible for the lack of knowledge about this phase. We discuss three of the more central factors next:

1. **Lack of reflective processes**: Because developers of technology do not tend to reflect on how the design process unfolds and what factors are having an impact on it, there is a general lack of reflective processes in the industry. Developers are primarily concerned with the day-to-day tasks at hand and less with the social, economic, cultural, and political factors that impact their design. Developers are also pressed with time, leaving little opportunity to engage in reflective processes in the same way that social scientists would. When, for example, Pasteur and his team developed the pasteurization process in the nineteenth century, they were not concerned with *how* they made the discovery; indeed, they had more pressing concerns, such as developing a more effective technique for food preservation that ultimately could help prevent illness and save lives (Latour, 1988).

2. **Retro-analysis**: Most accounts and analyses of how a technology was designed often become relevant, or have social significance, only after

the fact. For instance, Bruno Latour (1988) only examined the complexities of how the process of pasteurization was developed in the 1980s, almost 200 years after Pasteur and his team discovered the technique. Therefore, the majority of insights are gained through archives, memories of people involved, and any other paper or digital trail available. In addition, it is difficult to predict during the development phase which technologies will have a significant impact on society.

3. **Black box of design**: Many innovations occur behind closed doors as a result of competitive pressures, and hence it is difficult, or impossible, for researchers to obtain access to these developments as they unfold. This is particularly the case with research linked to the military, which often falls under the rubric "top secret." But this secrecy applies equally to innovations occurring in other industries because fears of spies are not a myth in R&D but often a reality. Former Google employee Paul Adams (2011) experienced the consequences of secrecy in the software development industry when he left Google to start working for Facebook and suddenly was no longer granted authorization to publish his book *Social Circles*. For Google, this work represented a major infringement of its intellectual property, as Google+ had not been released at the time and the company feared that information could leak to the public prior to its much anticipated release.

The three factors outlined are linked directly to our limited understanding of how design unfolds. This gap in our understanding has led to the development of a new field of study, often described as **science and technology studies (STS)** (Bauchspies, Restivo, & Croissant, 2006), which we covered in Chapter 3. Bijker, Hughes, and Pinch (1999) describe how STS has moved away from looking at technological design as a black box by following three important principles. The first is the **multiple factor principle**, which no longer stresses the relevance of the single inventor—the genius—but, rather, employs the idea of a coming together of individuals and economic, social, historical, and political factors. Latour (1988) stressed the relevance of this perspective when examining how Pasteur discovered the process of pasteurization. Prior to his analysis, full credit for the discovery had been attributed to Pasteur alone. Through Latour's in-depth analysis, he uncovered the relevance of co-workers, colleagues, farmers, and others in how the discovery came about. This case study demonstrates how innovations cannot always be attributed to a single individual but instead often come about as a result of multiple factors.

The second principle in describing how STS has moved away from looking at technological design as a black box is **mutual shaping**, which refers to a move away from the framework of technological determinism, which sees technology as the single most important factor leading toward social change.

Instead, in the mutual shaping framework, technology and society are both seen as forces that together influence and shape each other. To contrast these two different frameworks, we examine how social network sites, such as Facebook and Twitter, have influenced our friendship networks. From a technological determinism framework, we would argue that social network sites have increased the size of our friendship networks because they facilitate keeping in touch and they bridge communication gaps. Recent studies, however, suggest that technology does not directly cause social change as network size has not yet increased. Indeed, one study of Americans suggests that they have fewer friends than two decades ago (McPherson, Smith-Lovin, & Brashears, 2006). However, the more optimistic studies report a *moderate* increase in the number of friends (Wang & Wellman, 2010). Hence, networking technologies do not appear to completely change how we connect, even though they can modify many of the practices and norms around communication.

Finally, the **seamless Web of technology and society principle** outlines a move toward integrating political, economic, cultural, and social factors to explain how technological change occurs. That is, technological change is no longer examined on a single dimension but instead is viewed as a coming together of different societal factors in a seamless Web. From this viewpoint, the increased reliance on social media is not purely technological but reflects other trends. One such trend is migration. There has been an increase in people's mobility, creating a real need for people to be in touch over long distances. Similarly, as globalization continues, there is also an increased need for communication over long distances. This shows that it is not only technology driving communication—other trends in society occur simultaneously with technological change.

Systems Theory

In this section, we discuss the key premises of **systems theory**, a highly influential approach to the study of how technological and societal factors impact design. Systems theory examines how physical artifacts, social institutions, and the social context all interact in complex ways to influence design and cause social change (Hughes, 1983). In his influential book *Networks of Power: Electrification in Western Society, 1880–1930*, Thomas P. Hughes defines *systems theory* as the study of self-regulating social and natural systems that can influence their own behaviour through feedback loops. Bánáthy (1997) defines a **system** as "a configuration of parts connected and joined together by a web of relationships" (p. 22). Hughes developed systems theory with the aim of understanding the design and development process of technologies in the context of societal forces. For him, societal forces play a key role in the design and development process because technological changes are closely linked to the goals and interests of individuals, groups, and organizations. The case study Hughes employed to illustrate the use of systems theories for the purpose of studying technology was the development of the electric power grid. His primary goal in presenting this case study was to

> explain the change in configuration of electric power systems during the half-century between 1880 and 1930. Such change can be displayed in network diagrams, but the effort to explain the change involves consideration of many fields of human activity, including the technical, the scientific, the economic, the political, and the organizational. This is because power systems are cultural artifacts (Hughes, 1983, p. 2).

The systems approach allows researchers to simultaneously examine micro- and macro-level phenomena; for instance, design decisions made in Facebook's headquarters can be directly linked to trends unfolding in the wider society, such as the introduction of Google+, which has posed serious competition (Kirkpatrick, 2010). Hughes (1983) suggested that three stages characterize the design and development of a technology:

1. **Technology development**: In this phase, the technology is being invented and developed. The technology slowly takes shape, with inventors and entrepreneurs working on creating a prototype or demonstrating its utility. Often other players, such as managers and financiers, may also be indirectly involved in this phase.

2. **Technology transfer**: During the technology transfer stage, innovations are transmitted from one geographic area or social group to others. In this phase, agents of change play a central role in aiding in the transfer of technological know-how (see next chapter for details). Agents of change can include inventors, investors, managers, etc.—usually individuals with a vested interest in the spread of the innovation.

3. **System growth**: Examining the growth of a system is a methodologically challenging task. A good example of this process is the development of the personal computer (PC). The PC has experienced differential growth in its various components, including random-access memory (RAM), processor (CPU), screen quality, and storage capacity (hard disk). Central to the examination of system growth is an understanding of **reverse salient** or **salience**, which is an imbalance in the growth of a system's sub-components. The concept has been widely adopted in the social and computer sciences because "the socio-technical characteristic of the concept affords flexibility" (Dedehayir, 2009, p. 586). Often, one component will develop rapidly, leaving many other areas of development behind. For example, in the computer industry software developed rapidly, often

forcing PC hardware components to keep up with the requirements in speed and capacity (Dedehayir & Mäkinen, 2008). As a result, people had to update their computer every few years so that current software programs could be run efficiently.

Hence, a key task of developers is to work around irregularities and attempt to have the system fully functional as a coherent whole. In order to be able to tackle these reverse salients, developers must first redefine them as **critical problems**. Only then can potential solutions be proposed. A critical problem describes a challenge that requires a complex solution, often bringing to bear both technological innovation as well as institutional or social change. An example of a reverse salient that was redefined as a critical problem is the idea of a stand-alone PC. While early developments in computing were aimed at developing powerful processors, it became quickly apparent that the PC's value lay not in its capacity as a stand-alone devise but, rather, in its capability to network. This revelation required a complete rethinking of the challenges confronted by computing.

In the view of STS and systems theory, design is a central topic of inquiry that needs to address political, economic, social, and cultural factors. Without an understanding of how these factors play a role in the design of technology, technological progress is reduced to a purely technical process, ignoring how technology is a central component in our technological society. Next we discuss the concept of *technopole* and its close interplay with the global economy.

Technopoles: Centres of Innovation

The term the **information society** has been widely used to describe a shift in the economy from an era where production was at the centre of economic development to one where knowledge is considered a central asset (Fuchs, 2010). Several terms have been employed to describe this shift, including post-industrial society, Post-Fordism, knowledge society, the information revolution, and network society (Bell, 1973; Castells, 1996). In this new economic model, information and knowledge become key commodities, with capital and production still being relevant but as secondary industries affiliated with the information society.

As a result of the shift from an industrial society to an information society, the centres of power have also shifted. This new economy has led to rapid growth in industries linked to the creation of information products and services. Many terms have been used to describe the centres of innovation around the world, including *high-tech sectors, biotechnology,* and *tech hubs.* Castells and Hall (1994) use the term **technopole** as an umbrella concept and define it as "planned developments" whose function it is "to generate the basic materials of the informational economy" (p. 1). Technopoles have been created in many parts of the world and show a certain kind of homogeneity: they are complex socio-technical systems constituting a geographical area that is comprised of buildings, people, know-how, institutions, and corporations.

Castells and Hall (1994) view contemporary technopoles as arising from three interrelated processes:

1. **Information revolution**: A technological revolution built around information technologies, creating a need for the design of new digital tools, platforms, and content. As a result, a wide range of products and services are needed to sustain the information society. For this purpose, every day hundreds of new services are being offered online, including Amazon as a book retailer, Expedia as a travel broker, and Flickr as a photo-sharing site.

2. **Globalization**: The formation of a global economy that transcends national boundaries, governments, and laws. Globalization characterizes this new economy, which functions in real time across the globe, impacting all levels of society, including political, capital, labour, and management. The term **interdependence** is used to describe how the system is highly interconnected, with one component having trigger effects on other components located in different geographic regions. A good example of this is the impact of the US subprime mortgage crisis in 2007 on all aspects of the global economy, including the technology sector. We discuss Ireland on page 71 as an example of a country that has suffered from this economic downturn (Kirby, 2010).

3. **Informational production**: The development of new forms of economic production centred on information. *Informational* describes an economy where "productivity and competitiveness are increasingly based on the generation of new knowledge and on the access to, and processing of, appropriate information" (Castells & Hall, 1994, p. 3). This is a trend away from the agrarian and industrial models, where the sum of capital, labour, and raw materials created economic growth. In the new model, the outputs from science, technology, and information-related industries are the basis for economic expansion (Rao & Scaruffi, 2010).

These three processes lead toward the creation of technopoles as centres of innovation, where new technologies are developed and tested. In these highly competitive environments, technological innovation develops, as a result, from learning by doing instead of being based on off-the-shelf solutions (Castells & Hall, 1994). Therefore, technopoles have to be flexible and willing to constantly experiment with new ideas in order to remain competitive in a global market. These centres of innovation must, then, integrate seamlessly with service facilities in emerging economies as well as with production centres often located in developing economies, "creating a synergistic interaction between design, production and utilization" (p. 5). The design, testing, and prototyping of products may occur in North America, but the production process often takes place in Asia necessitating high degrees of co-operation and coordination among different firms. In Box 4.1, we examine Silicon Valley as a model of a technopole that unites highly skilled workers, a geographic area, social networks, and public institutions with corporations that have large sums of capital investment.

Before its contemporary inception, Silicon Valley had a historical research-oriented focus based on developments in electrical engineering at Stanford University (Rao & Scaruffi, 2010). As Castells and Hall (1994) explain, Frederick Terman, a professor of radio engineering at Stanford who later became dean of electrical engineering, was key to initiating and fomenting R&D initiatives in Silicon Valley. Terman used his influence to nurture gifted students, such as William Hewlett and David Packard, and provided them with funds to begin their own **startup companies**. Startups are newly created companies with a strong emphasis on R&D. They often rely on **venture capital firms** and **angel investors** to help them establish their operations. Hewlett and Packard formed HP and became one of the most well-known electronics companies in the world.

BOX 4.1 ☼ SILICON VALLEY: CENTRE OF INNOVATION

Silicon Valley is the first and perhaps most innovative technopole in the world. The area consists of a 40-mile by 10-mile (70-kilometre by 15-kilometre) radius in the San Francisco Bay Area, reaching from Palo Alto to San Jose (Castells & Hall, 1994; Henton et al., 2011; Rao & Scaruffi, 2010).

While in the 1950s little technological innovation was present in the Bay Area, a rapid expansion occurred in the 1970s, which continues to this day (Rao & Scaruffi, 2010). Table 4.1 shows a breakup of employment in the Bay Area between 1959 and 1985. The increase in jobs in such sectors as computing—from 0 in 1959 to 56,126 in 1985—is unprecedented. We can observe similar trends in the expansion of fields such as semiconductors, with an increase from 0 jobs in 1959 to 47,069 in 1985. A recent report on Silicon Valley shows that jobs continue to be plentiful, with companies trying to recruit engineers, designers, and computer scientists (Henton et al., 2011). The same report points toward another period of rapid expansion, documenting an increase from 2009 to 2010 of 12,300 jobs in the areas of computer systems design, employment services, and computer and electronic product manufacturing (Henton et al., 2011). Companies such as Apple, Google, and Facebook continue to hire, with Google alone adding 1,900 jobs in its recently completed 2011 fiscal first quarter (Swartz, 2011).

TABLE 4.1 Employment Structure in Silicon Valley, 1959–1985

Employment	SIC	1959	1965	1970	1975	1980	1985
Computers	3,573	0	0	8,938	19,902	52,738	56,126
Other office machines	375*	0	0	979	1,869	2,582	2,748
Communications	366	895	5,027	7,271	10,043	19,603	29,677
Semiconductors	3,674	0	4,164	12,290	18,786	34,453	47,069
Other electronic components	367*	4,295	4,619	14,174	11,622	25,472	23,731
Missiles/parts	372	0	0	2,274	0	0	750
Instruments	38	328	1,202	2,567	14,646	24,912	19,382
Drugs	283	0	282	0	750	1,976	1,954
Software/data processing	737	0	0	0	3,887	7,813	15,368
IC labs	7,391	118	2,193	1,978	1,642	3,856	6,133
Electronic wholesale	5,065	131	693	1,107	2,092	3,703	9,179
Computer wholesale	5,086	199	243	373	620	2,005	2,807
Total high-tech employment		5,966	18,423	51,951	85,859	179,113	214,924
Total manufacturing employment		61,305	88,038	131,613	154,126	256,437	272,332

Source: US Bureau of the Census – County Business Patterns.
Note: *SIC Codes 357 exclusive of 3573 and 3674, respectively.

Terman also had a vision for how Stanford could encourage entrepreneurship. He separated R&D from industry by creating the **Stanford Industrial Park**, which was built on university land but worked as a loosely connected entity. Terman invited talented individuals from the burgeoning electronics industry to Stanford Industrial Park with the idea of creating a community that fused technical research and scholarship. This is the optimization of the use of basic research to solve applied problems as discussed in Chapter 2.

Many companies were created as **spinoffs** of previous startups, which had disintegrated due to problematic working relationships. A spinoff company results from the split of a company into distinct businesses and is built upon the intellectual property, products, ideas, and tools that existed in the parent business (Rohrbeck, Döhler, & Arnold, 2009). For instance, William Shockley, co-inventor of the **transistor,** moved to California in the 1950s to start his own company. Shockley invited eight of the best graduates from the area, including Robert Noyce, who is the co-inventor of the **integrated circuit**. Shockley's poor business acumen and unlikable demeanour resulted in his eight engineers as a group starting up their own microelectronics company, Fairchild, which exclusively worked using silicon. Further splintering resulted in former Fairchild workers developing their own companies, for example, Intel. As much as 85 top companies have been created as spinoff companies from Fairchild. The fact that a new company was established every two weeks in the 1970s shows the rapid speed of economic growth in the area (Castells & Hall, 1994).

Companies in Silicon Valley also benefited from the desire for cutting-edge electronic equipment and devices by the US military and its related aerospace industries. By the 1970s, Silicon Valley companies had their industrial basis, as well as financial and research support, from military and commercial financial backers and nearby universities (Castells & Hall, 1994). This mix of diverse and influential players led to large growth in the technology industry.

Tall, cylindrical buildings are home to the Oracle Corporation headquarters in Silicon Valley, California.
Source: © iStockphoto.com/JasonDoiy.

The universities in Silicon Valley played an instrumental role in the area's unprecedented economic growth. For this growth to have taken place, an innovative and entrepreneurial workforce was needed, and the universities provided a constant supply of graduating students (as well as smart dropouts), who were eager to join what they perceived as a fun, exciting, and challenging work culture. These students were educated individuals—often with graduate degrees—who had the necessary skills for designing, manufacturing, and marketing high technologies. They were also highly motivated to participate in and benefit from this quickly expanding economy.

Another important factor in the unique expansion of Silicon Valley was the establishment of complex social and work networks. Employees frequently moved from one company to another as new opportunities emerged but, at the same time, maintained their informal social networks with prior colleagues. This facilitated the rapid exchange of information and the diffusion of new innovations creating new business, research, and development opportunities (Saxenian as cited in Castells & Hall, 1994). Members in groups featuring local computer enthusiasts included future key figures in companies such as Microsoft and Apple (Castells & Hall, 1994). For example, Steve Wozniak, the inventor of Apple, Bill Gates, the founder of Microsoft, and other young computer enthusiasts met regularly to exchange ideas and reflect upon new developments.

Silicon Valley was thus shaped by several factors that enabled it to become a key technopole. It featured an aggressively competitive **entrepreneurial spirit** within a highly educated and innovative workforce; an affluent culture able to invest in new, risky business ideas; and a firm technical base supported by high levels of public and private investment from education, military, and industrial sectors (Castells & Hall, 1994). In addition, Silicon Valley had a unique entrepreneurial culture that supported informal social networks and encouraged a combination of hard work and hard play.

Technopoles have emerged around the globe as governments are eager to capitalize on the growing technology industry. However, technopoles also bring risks to local industries (Kirby, 2002). Ireland is a prime example of an economy that showed rapid growth between 1995 and 2007 with many high-tech companies moving into the region. As a result, real estate developed quickly, jobs expanded, and the country experienced an economic boom that led to a nickname, **Celtic Tiger**, reflecting Ireland's rapid expansion into the high-tech industry. Figure 4.1 shows the ratio of exports based on foreign-owned companies versus indigenous companies from 1995 to 2004. This imbalance between foreign- and indigenous-owned companies is what has made Ireland's economy vulnerable to global changes because its economy heavily relies on foreign investment.

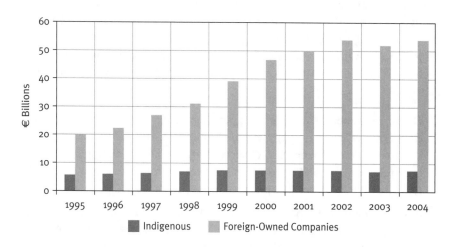

FIGURE 4.1 Irish Exports
Source: Forfás, Globalisation and the Knowledge Economy – the Case of Ireland, www.oecd.org/dataoecd/59/5/37563948.pdf.

In 2008, the country suddenly experienced a recession and in 2009 thousands went to the streets to protest as the situation continued to worsen (Kirby, 2010). Ireland's turnaround illustrates how technopoles are vulnerable to global trends and can dissipate quickly, leaving the local community with debt, unemployment, and poor living conditions.

Despite the risks associated with the development of technopoles, governments continue to heavily invest in the high-tech sector and to attract competitive companies through tax incentives, infrastructure, etc., in the hope of being part of the economic prosperity brought about by the informational economy. We discuss in Chapter 8 the so-called Internet dictator's dilemma that some nations face: on the one hand, they would like to profit from this new economy, but at the same time they want to continue controlling the information diffused to their citizens. The move toward digital societies is not just about economics; such a move also has social, political, and cultural consequences as the examples of Egypt, Libya, and Tunisia demonstrate, which we will discuss further in Chapter 8.

The Role of Research and Development (R&D)

In the previous section we discussed the relevance of technopoles to the economy. We focus now on the role of **research and development (R&D)** within industries, companies, and nations. No company can survive in a competitive market without R&D. Even companies who invest extensively in R&D can collapse if they do not focus on the right kind of technology. Both Kodak and Polaroid, for example, saw their business models and profits vanish as a result of inefficient R&D and slow uptake of digital photography. R&D is defined as consisting of "creative work undertaken on a systematic basis in order to increase the stock of knowledge, including knowledge of man, culture and society, and the use of this stock of knowledge to devise new applications" (OECD, 2009). The **Organisation for Economic Co-operation and Development (OECD)** includes three areas in its conceptualization of R&D and defines them as follows:

1. **Basic research**: Consists of experimental or theoretical investigation that is aimed at acquiring new understandings of the underlying foundations of phenomena and observable facts. What distinguishes basic research from applied or experimental research is that it is not intended to lead to any particular application or use in the field.

2. **Applied research**: Also constitutes original research conducted with the aim of acquiring new knowledge. In comparison to basic research, it is intended primarily to address or solve practical problems.

3. **Experimental development**: Systematic research based on existing knowledge that aims to develop new materials or products; to propose new processes, systems and services; or to modify processes already in place.

Even though all three areas contribute to the economy and are thought to provide direct economic returns, companies tend to invest primarily in applied research and experimental development, while universities and governments tend to support basic research. These differences in focus are a direct result of the expected economic returns of the three areas of R&D. Historically, there have been many different attempts to model the link between R&D and economic development. However, most of these theories were basic and did not provide much insight into the mechanisms that linked R&D to the economy: How does R&D yield gains? What mechanisms are in place to increase the economic returns coming from R&D?

Many scholars see Schumpeter as "providing the most comprehensive and provocative analysis since Marx of the economic development and social transformation of industrializing capitalism" (Elliott, 2004, p. vii).

Economic Development and the Creative Process

In his analysis of economic development, he identified the **creative process**—defined as the development of the economy through the production process—as a central activity of economic prosperity. Schumpeter then divides the creative process into three distinguishable stages: (1) invention, (2) innovation, and (3) imitation.

1. Invention

The stage of invention, which Schumpeter referred to as the **process of circular flow**, is characterized by general equilibrium because there is little change in the interrelation of economic factors. During this stage, the supply of goods is perfectly tailored to meet consumer demands and as a result no tensions or social problems arise in the social system. This stage has no innovators, leaders, or heroes because actors are only passive observers of a well-functioning economic system.

2. Innovation

For Schumpeter, during the innovation stage radical transformations occur, bringing about a fundamentally different social system. Schumpeter further proposed four kinds of changes as occurring during the innovation stage: (a) increases in salaries, (b) population growth, (c) changes in consumer tastes and choices, and (d) changes in how production occurs. While growth is an important aspect of the innovation stage, what is decisive is that change occurs not only quantitatively but also qualitatively. Schumpeter argued that even if you add "successively as many mail coaches as you please you will never get a railway thereby" (Schumpeter, 2004, p. 64, Footnote 1). Similarly, many high-tech companies have been created in the past two decades, but few of them have the profit margins, user-base, and influence of Google. This led Jeff Jarvis, from City University of New York, to propose in his book *What Would Google Do?* that Google simply had developed a radically different approach to packaging and offering information services to users. All the other newly created companies may be quantitatively different, but they are not fundamentally different and are lacking this element of innovativeness.

This proposal led Schumpeter to identify innovation as the single most important driver of economic prosperity, which he understood to be the "commercial or industrial application of something new—a new product process, or method of production; a new market or source of supply; a new form of commercial, business, or financial organization" (Elliott, 2004, p. xix). The consequence of innovation is a radical transformation in the economic structure of society. This transformation completely turns social structure upside down, and Schumpeter introduced the term **creative destruction** to summarize the social, economic, and cultural consequences that innovation brought about.

Innovation, while desirable, is not easy to accomplish. Schumpeter (2004) identified three key factors that impact innovation. First, innovation is based on new ideas and processes, making it difficult to establish the expected outcome and potential future revenue (Jarvis, 2009). In general, there is little information and knowledge about how processes will unfold, leading to great uncertainty. Second, because there is such great deal of uncertainty, it is difficult for investors to predict the **return on investment (ROI),** and hence there is apprehension in supporting the innovation (Miller, 2011). (Return on investment is the net return when the cost of investment and the gain from investment are taken into account.) Third, there is reluctance to accept change in society, making it difficult for innovators to convince others of the usefulness of their ideas. Even Alexander Graham Bell found it difficult to convince others of the usefulness of the telephone at first (Fischer, 1992).

These three obstacles to innovation become apparent when the **innovator-entrepreneur** is compared to the capitalist. The capitalist relies on detailed economic analyses, which show in predictable ways the revenues that will result from investment. By contrast, innovators are more heroic figures because of their willingness to take risks and their tendency to experiment with new forms of thought and action. As Elliot (2004) argues, at the centre of Schumpeter's social and economic analysis lies the figure of the innovator-entrepreneur, who through his vision drives technological change, which then will impact all aspects of society. The distinction between innovator-entrepreneur and capitalist became much more nuanced in 2011, with Google, for instance, investing approximately US$200 million in the acquisition of new startups (Miller, 2011).

3. Imitation

Schumpeter (2004) analyzed economic development and established that it does not evolve evenly but, instead, in **cyclical fluctuations**. By this, he meant that innovations occur suddenly, often clustered together. Why does economic development follow this kind of cyclical fluctuation? First, Schumpeter put forward the argument that these fluctuations are not indicators of a failure; quite the contrary, they are the normal way in which capitalist systems evolve and even prosper. Second, Schumpeter attributes the cyclical nature of the economy to the process of **imitation**. Once an innovation is diffused in society, others will learn about this innovation and want to adopt it, too. This process of imitation—both in the original industry as well as in secondary industries—results in economic growth. For example, an innovation in the car industry will lead other car manufacturers to follow and develop similar models. At the same time, other industries will develop products to support the car industry, such as gas, glass, roads, safety equipment, etc. Once the innovation is widespread and ceases to be new, capital investment diminishes as no additional ROI is expected. Further, "as an avalanche of consumer goods pours onto the market with dampening effects on prices; rising costs and interest rates squeeze profit margins: and the economy contracts: recession" (Elliott, 2004, p. xxvii). From this standpoint, recessions are a predictable and normal part of the cycle of economic development and to be expected as innovations reach their peak.

In summary, Schumpeter's economic and social analysis suggests that innovation is at the centre of economic development and closely linked to the structure of society. From this point of view, R&D is a centrepiece of economic development, and new technologies are forces of social, political, and economic change. In the next section, we examine how nations invest in R&D as a means to spur economic prosperity.

Global R&D

Why should nations care about R&D? And why should governments invest taxpayer money in R&D if that money could be invested in social programs, education, or health care? The expenditures linked to R&D accrued by companies, institutions, and nations are an indicator of their commitment to science, technology, and innovation. Based on Schumpeter's (2004) analysis, innovation can be directly linked to economic output. The higher the investment in terms of capital, people, and infrastructure that a nation makes in R&D, the greater the economic return. The **gross domestic expenditure on R&D (GERD)** is most frequently used for the purpose of comparing the investment in innovation that occurs across nations. The GERD index takes into consideration the total expenditure (current and capital) on R&D accrued by a nation over a one-year period, including expenses by companies, research institutes, universities, and government laboratories (OECD, 2009, Science and Technology Section).

Figure 4.2 shows the top countries' expenditures in R&D as a percentage of gross domestic product[1] (GDP) for 1998 and 2007, with Sweden being the country that spends the highest amount worldwide. Currently, only data for 2007 are available on the OECD website; however, because the OECD attempts to keep its statistics updated, more recent data should be available soon (see www.oecd.org).

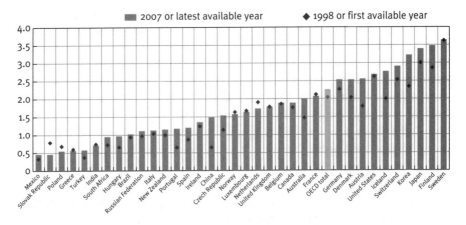

FIGURE 4.2 Gross Domestic Expenditure on R&D (as a Percentage of GDP)
Source: OECD (2009), OECD Factbook 2009: Economic, Environmental and Social Statistics, OECD Publishing, www.oecd-ilibrary.org/economics/oecd-factbook-2009_factbook-2009-en, http://puck.sourceoecd.org/vl=72896984/cl=11/nw=1/rpsv/factbook/07/01/01/07-01-01-g1.htm.

Important indicators of a nation's potential to innovate, in addition to the GERD, are the number of patents, peer-reviewed publications, and new products developed. While most nations are committed to providing support for R&D, the expenses linked to innovation need to be carefully weighed, for example with expenditures in the areas of health care, the military, and basic infrastructure. Because various groups in society have different interests, finding the right balance between economic and social interests is a challenge for most governments.

In addition to monetary investment in R&D, other initiatives have been developed to expedite and improve innovative processes. One recent global trend to encourage innovation has been to decentralize and spread R&D teams across the globe. Many large companies have adopted this approach, including Microsoft. Microsoft Research, the company's research arm, is devoted to carrying out both basic and applied research in the areas of computer science and software engineering. Instead of Microsoft Research physically being in a single location, it has three institutions in the United States (Redmond, Washington; Silicon Valley; and Cambridge, Massachusetts) and six additional locations (Bangalore, India; Cambridge, England; Cairo, Egypt; Beijing, China; Aachen, Germany; and Herzelia, Israel) (visit Microsoft Research's website at http://research.microsoft.com to see the worldwide locations on a map). The company's researchers usually work on local teams and have some involvement in global projects as well.

There are, however, R&D teams in other companies that are fully integrated across the globe. The purpose of these dispersed work structures is to benefit from the diversity of cultural backgrounds, expertise, and languages that a global team provides. In addition, a key advantage is the time gap between team members who work in different time zones. This allows team members to take on different shifts in developing the product: while part of the team sleeps, the new team takes over and continues development, helping reduce the time to completion considerably. These initiatives demonstrate the kinds of pressures experienced by R&D teams in a global, competitive market. For companies in the high-tech sector, managing innovation becomes a central feature for remaining competitive.

Understanding the Social Milieu of Software Development

Software development has become a central component of R&D. The relevance of software development, and its close link to content and users, can be seen in the high financial value of companies such as Facebook, Twitter, Flickr, and The Huffington Post. In 2011, AOL bought The Huffington Post, an American news website and aggregated blog, for $315 million. In 2006, Google purchased YouTube for $1.65 billion, even though it had not yet made a profit. Facebook, another company that has not yet made a profit, is now valued at $100 billion (Bilton & Rusli, 2012). The high financial value of these companies shows that code is not just a string of bits but a tool that facilitates complex social behaviours, such as informing, sharing, and collaborating at a large scale.

Considering the social and economic value of code, it is of great relevance to understand the nature of software development. Even though we may think that software development occurs in social isolation, it is in fact socially embedded. Carmel and Sawyer (1998) have developed a theoretical framework that emphasizes the social milieu in which software development occurs, and they distinguish between two different types of teams: **packaged software development** and **information systems development**. Packaged software development refers to software that is produced in large quantities and can be obtained off the shelf; Microsoft's Office suites are an example. This is in contrast to information systems development, which is customized software, such as Facebook, that is designed to meet the needs and requirements of a particular group of users. Further, Carmel and Sawyer's framework distinguishes four levels of analysis that are critical for understanding the work of these teams: (1) industry, (2) tasks, (3) cultural milieu, and (4) groups (Quan-Haase, 2009). Table 4.2 summarizes the key differences between packaged software development and information systems development for each of the levels of analysis.

To better understand the social milieu in which software development is embedded, we examine each level in more depth.

1. Understanding the Industry

The high-tech industry in which software development occurs is characterized by intense time-to-market pressure because companies compete to be the first to deliver software that includes the "latest" functionality at low prices (Dubé, 1998; Krishnan, 1998; Zachary, 1998). As indicated in Table 4.2, the time to market pressures are greatest in packaged software development teams because they are dependent on frequent and highly innovative releases. Currently, we observe these pressures in the tablet or e-book reader industry, where companies are in stark competition to acquire market share in this growing industry. For these companies, it is essential to acquire early market share and position themselves as a leader. Waterloo-based Research in Motion (RIM) continues to develop the PlayBook in a desperate move to keep up with Apple's iPads 1, 2, and 3, and with the Sony Reader.

TABLE 4.2 **Software Development Framework**

	Packaged Software Development	**Information Systems Group**
Industry	-Time-to-market pressures	-Cost pressures
	-Success measures: profit, market share	-Success measures: satisfaction, acceptance
Tasks	-Staff assigned to specific tasks	
	-User is distant and less involved in development	
	-Process is immature	-Staff assigned to specific projects
	-Software development via coordination	-User is involved and provides input
		-Process is more mature
		-Task accomplishment independent
Cultural Milieu	-Entrepreneurial	-More bureaucratic
	-Individualistic	-Less individualistic
	-Long work hours	-More set working hours
Groups	-Less likely to have matrix structure	-Matrix managed and project focused
	-Involved in entire development cycle	-People assigned to multiple projects
	-More cohesive, motivated, jelled	-Work together as needed
	-Opportunities for large financial rewards	-Salary-based
	-Large discrepancies in income	-Rely on formal specifications
	-Small/collocated	-Larger/dispersed

Source: Adapted from Carmel and Sawyer, 1998.

2. Understanding the Tasks

The primary task of software development teams is to write code. While all team members know how to program, each person is responsible for specific components of the software that require specialized expertise: the range of expertise often includes quality assurance, programming, marketing, design, and client services. In the information systems group, people are more closely integrated with the development of the product than is the case in packaged software. When software is developed for a client, usually that client will have very specific needs that the software has to meet, while packaged software is tailored toward a large user base.

3. Understanding the Cultural Milieus

The work culture of software development teams is characterized by highly individualistic work habits (Borsook, 2000). What this means is that developers have no predetermined work schedule, but their schedule and tasks are constantly changing. In the leadup to a release date, employees can work 50, 60, or even 80 hours per week, and the pressure at those times is high. Not all developers, however, have the same status within a team. The term **software cowboy** describes the high-performing developer, who is a "brilliant genius, who single handedly conceives and codes clever new systems in sweaty and sleepless weekends of nonstop programming" (Constantine, 1995, p. 48). There are always a few software cowboys in every team, who earn above-average incomes and excel in writing code and solving programming problems. To exemplify the role of these innovator-entrepreneurs in software developing, Box 4.2 discusses Mark Zuckerberg as the key developer in the initial stages of the development of Facebook.

Box 4.2 illustrates some important aspects of the cultural milieu of software development, in which it is essential for programmers to not only excel at writing code but also to grasp people's needs. Two key terms in computer science help bridge the gap between the writing of code and the requirements of users. *Object affordance* describes how physical objects can be designed in such a way that they facilitate behaviour (Norman, 1988). For instance, a door knob is a design that suggests to a person to open a door. The concept has been employed in human–computer interaction (HCI) to better understand how programs can be designed to elicit specific behaviours and facilitate the navigation of computer interfaces (Sears & Jacko, 2008). The term **social affordance** has been used to describe the "relationship between the properties of an object and the social characteristics of a group that enable particular kinds of interaction among members of that group" (Bradner, 2001; Bradner, Kellogg, & Erickson, 1999, p. 154). The term helps us understand how a simple change in the design

---

BOX 4.2 ❖ DESIGNING SOCIAL NETWORK SITES

In the early 2000s, a number of social network sites (SNSs) emerged on the Internet, including Six Degrees, hi5, and Friendster (boyd, 2006). Despite having a lot in common, it was because of small differences in the functionality of these SNSs that users migrated from one tool to another. Over time Facebook has become one of the most popular SNSs, with about 750 million active users in 2011 (Facebook, 2011). The movie *The Social Network* illustrates a number of unique characteristics of the cultural milieu of software development. Mark Zuckerberg's original intent is to create Facebook as an exclusive website for Harvard students (Kirkpatrick, 2010). He believes that Facebook will take off because it is based on Harvard's brand and supplements the already existing social structure of *exclusivity* that is so central to Harvard's culture of fraternities and sororities.

Zuckerberg, the inventor and key developer of Facebook, comes across in the movie as a genius developer who would certainly fall under the description of software cowboy, someone who spends long hours in front of the computer writing code (Kirkpatrick, 2010). What also becomes apparent when we analyze how Facebook's code developed is that Zuckerberg had a very good understanding of how social processes interfaced with the technological. Before the SNS took off, he was trying to find the relevant feature that would help make Facebook popular. He realized that "relationship status" was at the core of how users were interacting with the software. Although writing the code for including relationship status in the interface was straightforward, it was this social information that was relevant to this group of users. This shows how the writing of code for software cannot occur in isolation from people's needs and from relevant social processes. If software is not relevant for a specific user group, then people will simply not adopt it. So interface (usability) and content have to go hand in hand.

---

of a tool can have social repercussions. For instance, if a Facebook user changes her status from "in a relationship" to "single," this information becomes socially relevant as it will show up on her wall as well as in friends' news feeds. This update in relationship can trigger a series of events; for instance, friends and relatives may contact her to see what happened and how she is doing. Others will know that the relationship has terminated and provide emotional support. This update could also be interpreted as a signal that this person could potentially be available for new romantic relationships.

4. Understanding Team Structure

The existence of software cowboys does not preclude the team from showing high levels of co-operation and cohesion (Wysocki, 2006). Carmel (1995) labelled this type of highly cohesive working unit a **core team**. Team structures in software development vary considerably. A high level of communication and exchange between members of the team is necessary because of the interdependence of all components of the product. Moreover, consultation on design issues is also required because they affect the operability of the code.

Conclusions

The chapter addresses the often neglected topic of technological design, showing that its complexity results from the way that design unfolds. In the past, studies of design largely disregarded the influence of social factors, primarily focusing on the technology as the driving force. The field has moved away from understanding design as a technologically deterministic process, and several approaches have been developed within the field of STS to investigate the mutual shaping of social and technological factors in the process of design. These approaches provide new terminology and models to analyze how design unfolds and to examine its close interlink with social, economic, cultural, and historical factors.

The chapter introduces the concept of technopoles as an umbrella term to describe centres of innovation. Technopoles represent the engines of the information society because they comprise highly skilled workers, infrastructure, and capital investment for the purpose of designing and developing innovative products in the industry. However, not all regions of the world can afford to sustain technopoles—such as Silicon Valley—further increasing the gap between the haves and the have-nots. Even those regions that have developed technopoles are vulnerable to global pressures and changes in global markets, as the example of the rapid fall of the Celtic Tiger illustrates. We conclude with our analysis that in an information society, nations and markets are highly interdependent. Changes, even small ones, in one sector or one region can have social and economic consequences in other parts of the world.

Our analysis of GERD, the measure of investment in innovation, shows large discrepancies among nation-states. Sweden, Finland, Japan, and Korea are the largest investors in R&D in the world. Canada and the United States invest considerably in R&D and are positioned in places number 14 and 7, respectively. Mexico is the last country to be included in the chart with many other nations not included as they show almost no investment in R&D. On the one hand, there is large integration, mobility, and dependence among the key centres of innovation and world markets. On the other hand, the disparities between the haves and the have-nots, both between nations and within nations, are ever increasing. This necessitates future detailed analysis of technopoles and their social, economic, and political consequences.

Questions for Critical Thought

1. What are the key factors that have led to our lack of knowledge in the field of technology design?

2. Explain the concept of technopole using Castells and Hall's framework. Then discuss its economic, cultural, and social impact on local communities.

3. How does Schumpeter view the link between economic development and innovation? Do you think his theory continues to be relevant in the twenty-first century?

4. What are the key premises of object and social affordances? Discuss how this framework applies to the use of Facebook.

Suggested Readings

Borsook, P. (2000). *Cyberselfish: A critical romp through the terribly libertarian culture of high tech*. New York: Public Affairs. This interesting read provides an analysis of the high-tech culture of the late 1990s, in particular focusing on Silicon Valley.

Castells, M., & Hall, P.G. (1994). *Technopoles of the world: The making of twenty-first-century industrial complexes*. London: Routledge. This book provides an overview of the concept of technopoles and describes their social and economic relevance.

Kirkpatrick, D. *The Facebook effect: The inside story of the company that is connecting the world*. New York: Simon & Schuster. The book recounts the story of how Facebook developed into an influential startup, and discusses the impact of Facebook on social relations, marketing, and information dissemination.

Schumpeter, J.A. (2004). *The theory of economic development: An inquiry into profits, capital, credit, interest, and the business cycle* [1934]. New Brunswick, NJ: Transaction Publishers. This is the most influential social, economic, and political analysis of the link between economic development and the creative process.

Online Resources

SourceForge

http://sourceforge.net/

This site hosts thousands of software development projects created and developed by the community of open-source developers.

Microsoft's R&D global teams

http://research.microsoft.com/en-us/

This site provides an overview of the location and focus of Microsoft's R&D global teams.

5 | The Adoption and Diffusion of Technological Innovations

Learning Objectives

- To understand the process that underlies the adoption of technological innovations and the factors affecting this process.
- To understand the classic model of the diffusion of innovations.
- To examine key adopter groups, how the groups differ, and their characteristics.
- To investigate what social factors are most salient in the adoption of a diversity of technologies ranging from simple innovations such as boiling water to multi-media tools like the iPhone.
- To examine how social theory on the adoption and diffusion of technological innovations can be applied to market research.

Introduction

The study of technology typically focuses on those tools, techniques, and apparatuses that have been widely adopted in society. That is, we are most interested in technologies that are well-known and have made an impact on how we live. Although less attention is paid to technological adoption and diffusion, this area of research provides insights into a set of research questions that are central to society: Why is one tool adopted over another? What tools did users reject? What processes—at the micro and macro levels—are at play during the adoption and diffusion of an **innovation**? The study of adoption and diffusion shows that technologies are not a given but are a result of social, cultural, political, and historical factors.

This field of study is referred to as the **diffusion of innovations**. The field has grown exponentially: in 1962 there were 405 publications in the area, while in 1983 the number had increased to 3,085 (Rogers, 1983). In fact, a search conducted in 2010 for the terms *diffusion* and *innovation* on *Scholars Portal*[1] (limited to social sciences) yielded 54,619 entries, and these counts do not include the vast amount of publications in related fields, such as engineering, science, and medicine. One reason for this rapid growth is the pervasiveness of technology in our society and its impact on how we work, play, socialize, and communicate. We can readily see this pervasive impact when we examine the use of virtual worlds, such as *World of Warcraft*

(*WoW*) or *Second Life*, where users gather for work meetings, build virtual objects for fun, fight virtual enemies, and hang out to make new friends. Scholars no longer see technology adoption and diffusion as a purely technological matter but have come to see it instead as reflecting social processes as well. This chapter will provide an overview of the key concepts, theories, and research findings in the diffusion of innovations literature and discuss them in relation to the diffusion of specific technologies, such as water boiling, the QWERTY keyboard, and the iPhone.

Technological Innovations: The Process

The most influential researcher in the diffusion of innovations literature is Everett Rogers, a rural sociologist and communication scholar. His seminal work *Diffusion of Innovations* (2003) is the second most cited book in the social sciences and provides a comprehensive overview of diffusion studies.[2] In the book, he defines an *innovation* as "an idea, practice, or object that is perceived as new by an individual or another unit of adoption. An innovation presents an individual or an organization with a new alternative or alternatives, with new means of solving problems" (Rogers, 2003, p. xx). The term *innovation* is used interchangeably with *new technology* or *technological innovation*. **Adoption** is referred to as the decision to start using an innovation for a specific goal. While adoption takes place at an individual basis or organizational level, *diffusion* describes the process by which, over time, an innovation becomes adopted in a social group. Hence, diffusion is a macro-level phenomenon that is structural in nature.

Early models of the diffusion of innovations tended to place the innovation itself at the centre of the diffusion process. The assumption was that an innovation that was superior to previously established technologies would quickly spread and become widely used in society. Hence, diffusion was solely a matter of technological superiority. These models, however, did not capture the complexity of how diffusion occurs; adopters need to be considered as they are active participants in the decision-making process. These models also failed to account for the social, cultural, and political factors that affect adoption; individuals are members of social groups and the norms, beliefs, and attitudes of these groups will also shape adoption decisions.

Inventing and Copying Technology

Jared Diamond has been interested in the evolution and history of innovations. He argues that studying the origins of technological innovations is central to understanding social change in society. When looking at the history of tool use, for example, social groups may at first appear to be highly innovative, constantly developing creative solutions to existing problems. Based on the anthropological record, however, Diamond concludes that innovations are rarely developed locally; most innovations are taken from other social groups—this is referred to as the copying of technology. Social groups are less likely to invent new ways of dealing with problems as they are to copy existing solutions developed and tested elsewhere. Two key factors affect whether innovations are more likely to be invented locally or copied from other societies: **ease of invention** and **interconnectedness** (Diamond, 1997).

Innovations that are easy to invent are more likely to be developed locally, while innovations that are complex are usually copied. Diamond (1997) explains: "Some inventions arose straightforwardly from a handling of natural raw materials. Such inventions developed on many independent occasions in world history, at different places and times" (p. 254). Pottery is a good example of an invention that has several independent origins, probably because pottery results from the handling of a raw material, in this case clay. By contrast, the magnetic compass was invented only once in world history. Other inventions identified by Diamond that have been invented only once in the Old World and never in the New World include the water wheel, the rotary quern, and the windmill, which suggests that the spread of most technologies is a social process where one social group learns about a new technology from another.

Geographical location plays an important role in a society's adoption of technologies. Societies located in central areas, such as medieval Islam, for example, with access to both their own inventions and those of their neighbours were generally better equipped and more receptive to acquiring and adopting new technologies, in contrast to more geographically isolated regions and societies, such as Tasmania. Societies that were isolated needed to rely on their own ingenuity because they could not easily borrow ideas or inventions from other social groups. Conversely, the diffusion of some technologies was principally connected to a need to retain power over rivals, as in the Maori tribes' acquisition of muskets from European traders in the early nineteenth century New Zealand (Diamond, 1997).

While the study of technological innovations tends to focus on the spread of ideas and tools, there have been also instances in history where new technologies were abandoned, a process referred to as **technological regression** (Diamond, 1997). An example of technological regression was the arrival of guns in Japan in 1543 through Portuguese traders. Following this initial acquisition, the Japanese rapidly adopted the weapon, manufacturing their own guns and becoming one of the largest areas of gun ownership in the world. Within less than a century, however, guns completely disappeared from the island. There were three key mitigating socio-cultural factors against a complete adoption of the gun as a mechanism for war in the

Japanese context: (1) customs, (2) power, and (3) politics. The Japanese warrior class, the samurai, who held great political and cultural sway, were against the weapon. The samurai weapon of choice, the sword, was a powerful "class symbol," while ideas of samurai warfare were centred on single combats between swordsmen. They saw the gun, therefore, as an ungraceful weapon. The gun's popularity also receded due to its foreign origin, as goods and ideas produced outside Japan became increasingly despised as part of an anti-foreign backlash (Diamond, 1997).

The Classic Model of the Diffusion of Innovations

The most influential model of diffusion is Everett Rogers', which was first proposed in 1962 based on his seminal work in rural sociology on the spread of agricultural innovations. Rogers was keenly interested in how agricultural innovations found acceptance by farmers. He interviewed 200 farmers to find out what motivated them to either adopt an innovation or reject it. Unlike most of the previous work on the topic, Rogers focused on the social factors that influenced adoption. Based on his own work and subsequent studies in the field, he proposed a comprehensive model of the diffusion process.

For Rogers, the concepts of **uncertainty** and **information** are central to the diffusion process. Potential adopters are constantly dealing with high levels of uncertainty because they do not know whether or not the new innovation will yield the expected outcomes; hence, its adoption is coupled with risks. If they adopt and the innovation does not provide the expected personal and societal benefits, negative effects could ensue, including those at the economic, social, environmental, and societal level. The deployment of the atomic bomb in Hiroshima and Nagasaki during WWII is an example of a technology that had not been employed before but was tested in the context of war and yielded disastrous effects for humanity.

A more mundane example of uncertainty as described in Rogers' model of diffusion is the adoption of new hardware/software. On the one hand, with the release of the Apple iPad 2, consumers find it appealing to buy the new device to obtain the flexibility of downloading e-books, surfing the Web, checking email, and, most importantly, being able to annotate and edit text in a flexible manner. This multi-media device also provides a wide range of entertainment capabilities, including watching movies on demand, creating a complex music library, and participating in both single and multi-player online games. The iPad itself is also a symbol showing how tech-savvy a person is. On the other hand, there exists considerable uncertainty as to whether the new application will become widely used and accepted or whether another, more superior tool will be developed, making the iPad 2 obsolete shortly after its deployment. In addition, the quantity of e-books that will be available at cost-efficient prices on the iPad remains uncertain. The Nook and Kindle carry Project

Gutenberg's 1.8 million free books on top of 500,000 titles they make available to their users. Is it then worthwhile to spend C$599 for the iPad 2? Should users wait until the next **killer app** is available? Or should they buy the Kindle, Nook, or Sony Reader as better alternatives to the iPad 2?

Information, the second key concept in Rogers' model, becomes an important resource because it reduces uncertainty and helps potential adopters make a decision about a particular technological innovation's usefulness and efficiency. Information can be obtained from multiple sources, including the technology and its functioning, peers, the media, and individuals inside and outside one's social group. The quality and extent of the information that an adopter receives will determine the certainty with which that person can make a decision. The amount of information also affects the time it will take an individual to decide whether to adopt or to reject.

The Main Elements of the Diffusion of Innovations

Rogers defines *diffusion* as "the process by which an innovation is communicated through certain channels over time among the members of a social system" (2003, p. 5). In this definition, four main elements become prevalent in the diffusion of innovations: (1) the innovation, (2) communication channels, (3) time, and (4) a social system. We next discuss each element in the context of Rogers' model of the diffusion of innovations.

1. The Innovation

Innovations as defined earlier in this chapter consist of new ideas, practices, or objects linked to technology and hence encompass all elements of our proposed definition of technology—what was referred to in Chapter 1 as material substance, knowledge, technological practice, technique, and societal complexity. Not all innovations, however, are of the same type; indeed, the nature of innovations can be classified in a number of different ways, based on their complexity, target audience, etc. Some innovations take off quickly—blue jeans, a consumer good, took only five years to spread throughout the United States—while others diffuse slowly—the use of the metric system is still not widespread in Canada even though it was introduced in 1970 and had a dedicated government agency promoting it (the Metric Commission). If Canadians are asked today about their height and weight, most will reply in feet and pounds, respectively.

To distinguish between the wide range of innovations, Rogers identified five perceived characteristics of an innovation (see Figure 5.1): (1) relative advantage, (2) compatibility, (3) complexity, (4) triability, and (5) observability. These five characteristics of innovations help us predict the speed with which an innovation spreads and help us understand how the nature of the innovation itself affects the process. Each characteristic is discussed next.

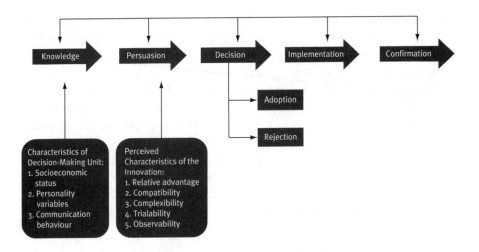

Figure 5.1 The Innovation-Decision Process
Source: Adapted from Rogers, E.M. (2003). *Diffusion of innovations* (5th ed.). New York: Free Press, p. 165.

First, **relative advantage** assesses the merits of an innovation in relation to the idea, practice, or object it is to replace. If potential adopters perceive that the innovation has added value, they will be more likely to adopt it. Early models tended to focus solely on the characteristics of an innovation as a means to assess its potential for adoption within a social group. That is, the models emphasized the "objective" advantages of a technology. This simple view is now obsolete because empirical studies have demonstrated that it is the "perceived" advantage of a technology that matters. An individual's perception is strongly influenced by social factors, including the social prestige associated with the innovation, the adoption rate of others in the social group, and information about the innovation received from communication channels.

Second, **compatibility** refers to an innovation's fit with a social group's existing norms, values, and attitudes. Only in those instances where a good fit exists will an innovation be adopted. For example, birth control continues to be rejected in a large number of communities across Latin America because they view birth control as inconsistent with their religious and cultural beliefs.[3]

Third, **complexity** describes the level of proficiency needed to comprehend the workings of a technology and its ease of use. People will be more likely to adopt technologies that are transparent in their function and benefits than those that are difficult to operate. Photography, for example, diffused slowly in society because people were not sure at first how cameras worked and what the consequences would be of depicting their image. Even today, in rural areas of developing countries people refuse to be photographed because they fear that the camera could take away their soul.

Fourth, innovations that users can test prior to implementation are more likely to be adopted than those that cannot be tried. **Triability** provides potential adopters with the possibility to reduce the uncertainty associated with the innovation and gather evidence about its value and associated potential risks through prior experience.

Last, Rogers identified the characteristic of **observability**, which refers to the visibility of the innovation itself and its benefits to other members of the social group. If others can see the innovation and how it works, they are much more likely to discuss it with each other and thereby obtain valuable information that can help them with the decision-making process.

2. Communication Channels

Central to the process of diffusion of innovations is the communication that occurs between individuals. Communication channels are the means by which the exchange of information occurs. The most commonly utilized forms of communication for the purpose of diffusing an innovation are the mass media, including television, radio, newspapers, and the Internet. While mass media are the most efficient means of reaching a large audience, they are not as effective at persuading a potential adopter as are interpersonal channels. Communication among peers is effective because peers can provide first-hand accounts about their experience with an innovation. Individuals are more likely to trust these accounts than messages coming from the mass media, which are often representing the interests of vendors.

In the literature, two additional reasons have been identified for the strong influence of personal relationships in comparison to that of mass media. First, a large number of individuals are not exposed to mass media because they do not read a newspaper on a regular basis or watch news on television. Thus, personal relationships have a greater reach with people often chatting about technologies informally. These discussions with friends and family are an important component in the decision-making process. Second, individuals who are less interested in technologies are more likely to discuss an innovation informally with a friend than to obtain information from formal sources, such as newspapers or the radio. Their general involvement with technology is low and hence most information they will obtain about the innovation will occur accidentally through encounters with other adopters in the social circle.

3. Time

In the social sciences, time is usually a variable that is of little relevance. Most studies completely ignore time because it is a part of all phenomena and therefore does not require much attention. Research in the diffusion of innovations, in contrast, is unique in the social sciences in terms of its emphasis on time, which is a central variable for examining when individuals first become aware of an innovation. A technology can be categorized into distinct adopter categories based on when individuals adopt that technology. Time also helps to explain an innovation's rate of adoption in a particular social system, which is measured as the number of adopters within a given time period.

4. Social System

Rogers (2003) defines the social system as "a set of interrelated units that are engaged in joint problem solving to accomplish a common goal" (p. 476). The units are individuals, groups, or organizations who are interested in acquiring information about a technological innovation in order to make a decision about whether to adopt or to reject that innovation. Examining the diffusion process within a social system is necessary because the structure of the social system as well as its values, norms, and beliefs have a major impact on the diffusion process.

The Innovation-Decision Process

After reviewing numerous adoption studies, Rogers was struck by the finding that the rate of adoption was similar across a wide range of innovations. When frequency of adoption was plotted on a graph by time, he discovered that the rate of adoption formed an s-shaped curve, which Rogers termed **s-shaped curve of adoption**. Figure 5.2 shows how few individuals adopt an innovation early, the majority adopt an innovation at the midpoint of the cycle, and a few adopt the innovation very late. The moment in time where the majority of individuals in a social group adopt an innovation is referred to as the point where the s-shaped diffusion curve takes off and represents a critical turning point for the diffusion of any innovation because it is at this point that it has been widely adopted in society.

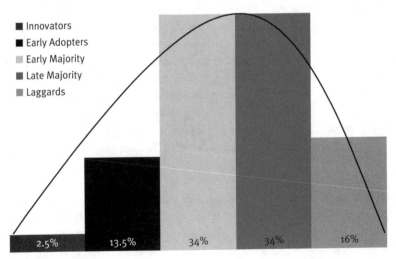

■ Innovators
■ Early Adopters
▪ Early Majority
■ Late Majority
▪ Laggards

2.5%   13.5%   34%   34%   16%

FIGURE 5.2 Adoption Categories
Source: Adapted from Rogers, E.M. (2003). *Diffusion of Innovations* (5th ed.). New York: Free Press, p. 247.

To describe what occurs during the adoption process, Rogers (2003) examined the innovation-decision process and identified five distinct stages: (1) knowledge, (2) persuasion, (3) decision, (4) implementation, and (5) confirmation. Figure 5.1 (on page 90) depicts the various stages that an individual goes through when making a decision about the adoption of an innovation.

First, in the **knowledge stage** an individual learns about an innovation for the first time and obtains information about how it operates. Coleman, Katz, & Menzel (1966) found that information about an innovation was obtained quite by accident through conversations with peers and interactions with sales personnel, and by exposure to the media. Individuals vary in terms of how receptive they are to information about technological innovations. Those individuals who have a need to solve a specific problem or who want to accomplish a particular task are more receptive to obtaining information related to their need. Hence, individuals seek out messages that are in agreement with their interests, needs, and attitudes. The process of seeking out relevant information is referred to as selective exposure and shows how individuals may be exposed to the same information but evaluate its relevance in different ways based on their needs.

Different kinds of knowledge are relevant in the knowledge stage. **Awareness knowledge** initiates the process by making an individual aware of the techology's existence and its potential adoption. From the awareness of the innovation emerges a need for **how-to knowledge**, which explains how the innovation is used properly and in what settings it can be employed beneficially. Finally, **principle knowledge** encompasses an understanding of the mechanisms that lie behind an innovation. For example, understanding the usefulness of and need for sanitizers is based on mechanisms outlined in germ theory.

Second, in the **persuasion stage** potential adopters continue to seek out information about an innovation, albeit now in a very active manner. Through such additional information, adopters develop either positive or negative emotions toward the innovation; these attitudes combined with the knowledge acquired earlier will help them to make a decision.

Third, the **decision stage** in the innovation-decision process describes the activities that lead toward adoption or rejection of an innovation. As defined earlier in this chapter, *adoption* refers to the decision to use an innovation. Rejection, by contrast, is the decision not to use a new technology. The decision-making process is often described as clear-cut: either an innovation is adopted or it is rejected. In reality, the boundaries between adoption and rejection are blurred, with many potential adopters first partially adopting an innovation by using it on a trial basis, defined previously as *triability*.[4] Once they have tested the innovation, they make a conclusive adoption decision. When it is not possible to test the innovation first, then they will seek the opinion of peers who have had first-hand experience with it.

Rogers terms this *trial by others* because the opinion of other users will help inform a person's adoption decision.

Fourth, in the **implementation stage** an individual starts using the innovation, often running into technical problems. Most diffusion research has tended to focus on changing attitudes or adoption behaviours, and has placed less emphasis on implementation itself and the social consequences of adoption. At this stage, information seeking continues to be central as the adopter troubleshoots the encountered obstacles. If users find it difficult to solve the problems encountered, they may stop using the new technology and return to what they are familiar with.

Fifth, following the implementation of a new tool, adopters continue to seek out information about the innovation to confirm that they made the right decision by adopting the innovation. This stage is referred to as the **confirmation stage** and unfolds over a lengthy period. If information arises that disconfirms the usefulness of the innovation, adopters will either try to seek out further information or may consider discontinuing use of the innovation, particularly if using the innovation has negative consequences that are confirmed by peers as well. This has been referred to as **discontinuance** and describes the rejection of an innovation after earlier use. If researchers fail to consider discontinuance, conclusions about an innovation's acceptance rates may be overestimated. Several factors impact discontinuance, including perceived characteristics of the innovation, competing innovations, evaluations of peers, and the fit with the social system's norms.

While the hype we encounter in the media would suggest that new technologies are adopted at a rapid pace, this is not the case. Not only that, even new technologies that are advantageous often do not diffuse in society. Rogers (2003) writes, "Many technologists think that advantageous innovations will sell themselves, that the obvious benefits of a new idea will be widely realized by potential adopters, and that the innovation will therefore diffuse rapidly. Unfortunately, this is very seldom the case. Most innovations, in fact, diffuse at a surprisingly slow rate" (p. 7), as is illustrated in Box 5.1 with the example of the nondiffusion of the Dvorak keyboard.

BOX 5.1 ❖ **DIFFUSION AND ADOPTION OF THE QWERTY KEYBOARD**

Early theories of the diffusion and adoption of technologies focused on the characteristics of the technology itself as a means for understanding individuals' willingness to embrace a tool. Many examples, however, have shown that the adoption process is neither linear nor unidirectional; instead, many social factors play a role in how technologies are diffused in society. An example of **nondiffusion** is the Dvorak keyboard. The standard keyboard, which most of us use several times daily, is the QWERTY keyboard, which was invented in 1873 and is named after the letters formed on the keyboard's upper left row. The main reason for introducing the QWERTY keyboard was to slow down typists. In the late eighteenth century, typewriters consisted of type bars that had to be stroked. When typists stroked two letters quickly, they would often get caught and the machine would require repair. To avoid the rapid typing and entangling of type bars, Christopher Latham Sholes designed a keyboard that positioned the letters in such a way that it was awkward to type the most frequently used sequences, thereby slowing down typists considerably while not diminishing the machine's satisfactory performance.

In 1932, August Dvorak, at the University of Washington, designed a keyboard that would speed up typing again. He created an arrangement in which the most frequently used letters—*A, O, E, U, I, D, H, T, N*, and *S*—were located across the top row of the typewriter. Moreover, the letters were arranged in such a way that the right hand, the stronger hand, would be performing more of the work, instead of the left hand, as is the case with the QWERTY keyboard. The QWERTY keyboard has a number of inefficiencies in addition to its slow speed, namely fatigue, higher error rate, and disorientation on the keyboard. Despite the many advantages of the Dvorak keyboard over the standard keyboard, it failed to diffuse in society, even though the American National Standards Institute and the Computer and Business Equipment Manufacturers Association approved it. The main reasons for the lack of diffusion seem to be associated with existing interests on the part of the manufacturers and the sales industry as well as the habits of typists themselves (Diamond, 1997). As Paul David (1985) has noted, Dvorak lost out to QWERTY due to factors of "technical interrelatedness, economics of scale, and quasi-irreversibility of investment," which enabled QWERTY to become "locked-in" (p. 334).

Source: Rogers, E.M. (2003). *Diffusion of innovations* (5th ed.). New York: Free Press, pp. 9–10.

The persistence of the QWERTY keyboard into the twenty-first century, even though it is inferior to the Dvorak keyboard in so many ways, shows how adoption and diffusion are often unrelated to the efficiency, benefits, and utility of technologies themselves but instead are closely associated with the social system and its complexities. Box 5.1 shows that many social factors play a role in the adoption and diffusion of technologies, even in developed societies. This is an important consideration because we often tend to ascribe to developing nations a lack of rational choice while assuming that individuals in developed nations make innovation decisions based on efficiency alone. The QWERTY example demonstrates that when it comes to the adoption and diffusion of innovations many factors need to be considered simultaneously, not all of which are rational. One of the key factors affecting the diffusion of innovations is the role of change agents, which we discuss in the next section.

The Role of Change Agents

Even though **change agents** are not depicted in the model in Figure 5.1, they are instrumental in all stages, with their most central role being to make potential users aware of a technological innovation and to provide them with information that will help them to make an informed decision about whether to adopt or to reject the innovation. Box 5.2 provides an example of how a change agent attempts to introduce an innovation into a social group. The example is of a failed diffusion campaign and highlights the challenges a change agent confronts and the large number of social, economic, and cultural factors that need to be considered when examining the role of change agents. Once again, we learn that the diffusion of technological innovations is not a linear, rational process where the innovations themselves are at the centre of diffusion but, rather, a complex social process taking place at many levels of analysis.

Box 5.2 ❊ **The Innovation of Water Boiling as Introduced by a Change Agent**

The Ministry of Health in Peru set out to introduce innovations geared toward health improvement and disease prevention in rural villages in Peru. A key innovation that the ministry was introducing as part of this program was water boiling to reduce the spread of illnesses, such as typhoid. While the innovations had clear health benefits, they failed to obtain widespread acceptance. In a village called Los Molinos in the coastal region of Peru, a total of 11 of the 200 families who lived there adopted the innovation.

The key sources of water in the village are a seasonal irrigation ditch, a spring, and a public well. Primarily children, and sometimes women, are given the task of fetching water. Water boiling was introduced by the local health worker, Nelida, who in this case is the change agent. The goal was to introduce homemakers in Los Molinos to the innovation of water boiling and to then persuade them to adopt water boiling as a routine behaviour to prevent the spread of disease. The program ran over a two-year period and Nelida spent considerable time with the families in the village.

Nelida organized a series of public talks given by a medical doctor to inform the villagers of the need to boil water regularly. In addition, she visited every family in the village several times to talk to them about water boiling and its advantages. She gave special attention to 21 families and visited each around 18 times. Of these 21 families, 11 adopted water boiling into their daily habits, while no other families in the village adopted water boiling. How can we understand this lack of acceptance in view of the clear benefits for the community?

A number of factors led to the demise of the campaign. First, water boiling was in opposition to cultural beliefs and local values. Hot water is only given to sick individuals in the community; healthy individuals will only drink cold water. This was incongruous with the message Nelida was attempting to convey, namely that boiled water would lead toward improved health.

A second important factor was the approach Nelida took. She worked with several women who were considered outsiders in the community. They were more open to Nelida's message and willing to adopt the new innovation because as outsiders they were seeking her approval and attention. With these women their adoption decision had less to do with embracing germ theory and understanding the advantages of the innovation and more to do with their position in the village's social network as outsiders. By contrast, Nelida did not approach villagers who occupied central positions in the village's social network, even though they could potentially have had greater influence on the adoption process.

Third, many women in Los Molinos were suspicious of Nelida. They felt that Nelida wanted to inspect their homes to determine their cleanliness and felt their privacy was invaded by her presence—they referred to her as "dirt inspector." These homemakers also rejected her because they felt Nelida represented the values of Peru's middle class and hence they could not easily identify with her. Rogers (2003), in his final analysis of the Los Molinos case, concludes that the change agent, in this case Nelida, "was too 'innovation-oriented' and not 'client-oriented' enough. Unable to put herself in the role of the village housewives, her attempt at persuasion failed to reach her clients because the message was not suited to their needs" (p. 5).

Source: Wellin, E. (1955). Water boiling in a Peruvian town. In B.D. Paul (Ed.), *Health, Culture and Community*. New York: Russell Sage Foundation.

The example of Los Molinos portrays change agents as central players in the diffusion of innovations process. In Los Molinos, the role played by the change agent was complex and directly linked to the socio-cultural context of the village. The change agent failed in promoting the innovation, despite it providing many advantages to villagers. In the next section, we continue exploring the diffusion process by looking at different adopter groups.

Classifying Adoption Categories

Research consistently shows that members of a social group adopt innovations at a different pace, with some members adopting an innovation early in the process and others adopting it much later. Early researchers realized the importance of identifying and categorizing adopter groups, but could not agree on terminology and method of categorization. In 1962, as mentioned, Rogers proposed the s-shaped curve of adoption to classify adopters, and this quickly became the dominant method of categorization.

The relevance of the s-curve of adoption to classifying adopter categories becomes evident when the cumulative rate of adoption is plotted instead of the absolute frequency by time. This curve has the shape of the normal distribution, often also referred to as bell curve, with most individuals adopting halfway through the adoption period. Rogers employed this bell-shaped curve as a means for categorizing adopters based on when they adopted a new innovation relative to others in the social system. Five key types of adopters were identified (see Figure 5.2 on page 92):

1. **Innovators**: These individuals constantly seek out new ideas and information, and this thirst for newness often leads them to expand their social circles—which is why they are often referred to as cosmopolitans. Their role is that of importing new ideas into their social group from the boundaries; that is, they are the **gatekeepers**. Because innovations are risky, they need to be able to cope with the uncertainty associated with adopting new tools, devices, and procedures.

2. **Early adopters**: Individuals in this group play the most central role in the diffusion of innovations process. They are well respected among their social network and serve as role models. Others in their social group will come to them for advice on the usefulness of technological innovations. As a result, they have been termed **opinion leaders**: individuals whose opinions influence the attitudes and behaviours of others in the social group. Unlike innovators, who are cosmopolitans, early adopters are similar to others in their social group and have ties to people in the community.

3. **Early majority**: These are individuals that are a step ahead of the average adopter. Before adopting an innovation, they go through a careful deliberation process where they consider the pros and cons of new ideas vis-à-vis the status quo. In comparison to innovators and early adopters, they have no gatekeeping or opinion leadership role in their social group. Instead, they serve as **interconnectors**; that is, they serve as a bridge between early and late adopters, providing relevant information about the innovation to the late adopters and often convincing them of the usefulness of the adoption.

4. **Late majority**: Members of this group adopt only after the average adopter has started using the innovation. They wait for everyone else to adopt first to reduce the uncertainty associated with adopting innovations. For these individuals, adoption occurs as a result of economic forces combined with pressure from their peers. The late majority are hesitant to adopt and skeptical of anything new; they would rather stick to the old ways of doing things.

5. **Laggards**: This group represents the traditional individuals who are the last in a social system to adopt an innovation. Laggards are often isolated, or are more locally oriented and connected to others with similar traditional values. Ironically, by the time laggards adopt a technology, new innovations have often already superseded the technology in question. Hence, when laggards adopt a technology, all members of a social system have adopted the innovation, and a state of market saturation can be observed.

Socio-economic characteristics perform a crucial function in shaping individuals' ability or willingness to adopt new concepts, technologies, and innovations. Early adopters, for example, have been shown to be wealthier and better educated than the laggards. Additionally, early adopters have shown an ability to absorb abstract concepts, accept change, seek information, and be well connected with both communication channels and peers, more so than laggards. Due to their increased financial resources, early adopters are more willing and able to take on the risks associated with new innovations, such as investing in a new idea or technology, which invariably may never be diffused into the mainstream.

Rogers' (2003) adoption categories have important implications for how we understand the spread of new technologies in society. The five groups represent differences in how individuals in society approach new technologies and in their willingness to adopt. The categories show that innovations take time to spread, with innovators and early adopters employing new technologies first and others in society slowly following their example. Hence, time is an important factor when examining the spread of innovations as well as the innovation-decision process. To show the applicability of the concepts, we will discuss several current examples in the remainder of the chapter.

Marketing Relations with Early Adopters

This chapter has discussed so far Rogers' classic model of the diffusion of innovations in detail. Even though the model was proposed in the 1960s, it has had a continued impact on disciplines such as sociology, communications, technology adoption studies, and science and technology studies (STS). In this section, we will examine how social theory on the adoption and diffusion of technological innovations can be applied to market research. This is a topic that is of great relevance to those developing digital tools, software, hardware, or any new technology. We will discuss a series of on-the-ground examples to illustrate the continued relevance of Rogers' classic model.

The literature on digital innovations is one area that shows extensive use of Rogers' social theories. His work has widespread acceptability and applies to many market analyses of digital artifacts. Trendsspotting.com tech analyst Taly Weiss discusses the role of innovators and early adopters in online marketing as follows: "[I]t's understanding the role they play and how they are to be approached. Getting it wrong can cost companies revenue. But getting it right can propel a product or brand to the front of the pack" (Rich, 2010). Market analysts continue to worry about how innovators and early adopters behave because they understand that having a critical mass is central to the diffusion of any technological innovation, digital or otherwise.

Laura Rich, a technology analyst from Yahoo.com, provides a compelling examination of how Apple and Google have dealt with early adopters. The case studies show that Apple neglected early adopters, particularly in helping its products diffuse to the wider population, and had to quickly rethink its marketing strategy to keep this group's support. When, in 2007, the iPhone was released in the United States, its sticker price was US$599. Many early adopters wanted to be the first to own an iPhone and because of the demand had to line up at stores for hours to obtain one. For these early adopters, being the first to own an iPhone was important and lining up was only a small hassle when considering the reward of owning the desired gadget. These are not average users of cellphones but a sub-set of people who feel strongly about Apple products and for whom the iPhone is more than just a cellphone. These early adopters of Apple products—who call themselves "fanboys"/"fangirls"—felt let down by Apple when only a few months later the price dropped to as little as US$399. On the Web and through traditional media, they expressed their outrage and disappointment with Apple. Steve Jobs, Apple's former CEO and one of the most influential players in the technology sector, realized the mistake Apple had made by offending their most devoted and trusted users. In a public apology he directly addressed early adopters:

> [E]ven though we are making the right decision to lower the price of iPhone, and even though the technology road is bumpy, we need to do a better job taking care of our early iPhone customers as we aggressively go after new ones with a lower price. Our early customers trusted us, and we must live up to that trust with our actions in moments like these. We want to do the right thing for our valued iPhone customers. We apologize for disappointing some of you, and we are doing our best to live up to your high expectations of Apple. (Jobs, n.d.)

How could Apple, a key leader in the sector, have disregarded its early adopters? Despite their central role in the diffusion process, early adopters work behind the scenes: they are not paid by companies, and they do not play any official role. As Figure 5.2 shows, however, they are part of the bumpy beginning that many innovations need to overcome. And despite not having a formal role, early adopters are central to the process for three main reasons. First, the work of early adopters is done most of the time through their personal connections with family, friends, neighbours, and co-workers. It is in this mentor role that their influence on the adoption process plays itself out. Second, they often disseminate the information about an innovation by showing strong endorsement. For example, early iPhone adopters used their devices in public spaces, attracting the attention of other potential adopters, thereby indirectly contributing to publicity by creating product awareness as well as endorsement. Third, early adopters have gone through the innovation-decision stages. They can then share information they have gathered about the product and the attitude they developed with potential adopters.

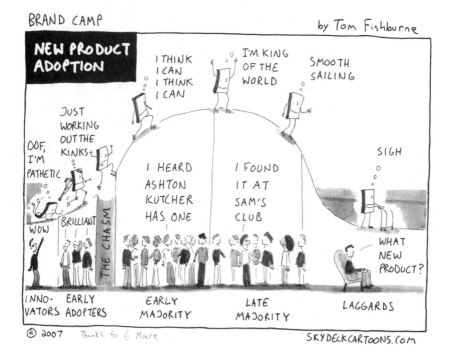

"Marketoonist." Tom Fishburne puts a humorous twist on the s-shaped adoption curve. Source: © Tom Fishburne, Skydeckcartoons.com.

Another great case study of how a new media product has diffused with the help of early adopters is the mainstreaming of YouTube—both its consumer base as well as its producer base. YouTube gained popularity as a result of early adopters' forwarding links to YouTube videos via email and instant messaging. These early adopters would often send mass emails to friends, making them aware of new, meaningful, bizarre, or comical videos.

Two YouTube videos were instrumental in the spread of the service into mainstream: "Lazy Sunday" and "Numa Numa." In December 2005, early adopters started forwarding the link to a *Saturday Night Live* sketch titled "Lazy Sunday"; it quickly became viral, introducing people to YouTube for the first time. The video featured Chris Parnell and Andy Samberg dancing and singing a farcical hip-hop song about cannabis. A second video that was instrumental in making YouTube mainstream and encouraging users to become content producers was "Numa Numa," which shows Gary Brolsma performing the song "Dragostea Din Tei" by the Moldovan band O-Zone. The original "Numa Numa" video has been seen 700,000,000 times worldwide—the second-most-viewed viral video after the video of the *Star Wars* kid, showing Ghyslain Raza swinging a golf-ball retriever (BBC, 2006). These videos showed the ease with which YouTube could be used and the potential for users to create their own videos and share them on YouTube, giving YouTube the much needed exposure to become widely known and utilized in society. As of 2011, YouTube has slightly more than 500 million unique users every month, which represents about one out of three Internet users globally (Sacks, 2011).

In this section, we have looked at how Rogers' classical model of the diffusion of innovations has been applied to market analysis. Analysts pay a lot of attention to their early adopters because they realize that having a critical mass is central for the adoption of any innovation, and in particular for the adoption of digital tools and Web 2.0 sites and services. Not paying careful attention to early adopters can really backfire, as companies may lose their most valued customer, as the example of Apple demonstrated. Early adopters play a number of key roles in the diffusion process by creating product awareness, endorsing the product, and spreading the message via word of mouth, which often happens online on Facebook, Twitter, and Flickr.

Conclusions

This chapter provides an overview of methods and processes affecting technological diffusion in both historical and contemporary terms. The value placed on newly acquired ideas, practices, and objects, which Rogers defined as innovations, can differ radically between societies, spatial areas, and historical eras. Those who see a tool's distinctive benefit may consider adopting, borrowing, or re-calibrating this technology to the specifications of their social, economic, or cultural needs. However, acceptance is not a given. Considerations such as uncertainty, utility, cultural adaptability, and economic resources may guide this decision-making process, as described in Rogers' model of the innovation-decision process. Ultimately, members of a social group may reject a technology or alter their perceptions of its usefulness, resulting in the abandonment of an object or idea in spite of its benefits—which is described as technological regression.

In a high-tech society, modern consumers, organizations, and governments are faced with similar concerns, questions, doubts, and analyses. For every technological success, there have been notorious failures. For example, portable electronic devices, such as the iPod or BlackBerry, have become radically integrated into society, bringing with them a new colloquial language and set of expectations for managing, organizing, and conducting one's personal and professional life. In contrast, many recent technologies— e.g., DIVX disc, the Apple Newton, and the Segway—have been rejected primarily due to factors such as cost, perceived lack of usefulness, or incompatibility with other accepted technologies. Similarly, as with the Dvorak keyboard, there are examples of contemporary technologies, such as HD DVDs, that were unable to diffuse within society due to the increased acceptance and industry support for a competing technology.

As our world rapidly changes, we must understand the social processes underlying technological diffusion. The existence of technological divides among adopter categories provides evidence that the rapid disposal of aging technologies in favour of new-fangled gadgets may not always be the best course of action. Additionally, a technology deemed to be inadequate, ancient, or inferior in one culture may prove to be relevant and necessary to another social group.

Questions for Critical Thought

1. How are the concepts of uncertainty and information linked in the innovation-decision process?

2. Define the process of technological regression and discuss what sociological variables have an impact on it.

3. Using QWERTY as an example, discuss the concept of nondiffusion. What are the types of socio-economic or cultural variables that can underpin nondiffusion?

4. Discuss the role of change agents in the innovation-decision process. What factors must change agents consider when introducing an innovation?

5. Compare and contrast innovators and early adopters. Which of the two adopter categories is more important for the diffusion of innovations, and why?

6. What makes Rogers' diffusion of innovations model so relevant for the study of social media? Discuss using Facebook as an example.

Suggested Readings ....

Choia, H., Kimb, S.-H., & Lee, J. (2010). Role of network structure and network effects in diffusion of innovations. *Industrial Marketing Management, 39*(1), 170–177. Investigates why an innovation sometimes diffuses throughout society, while other times it diffuses only to sub-groups.

Diamond, J.M. (2005). *Guns, germs, and steel: The fates of human societies.* New York: Norton. Diamond's award-winning book analyzes the various structures, values, networks, and properties that have shaped the acceptance and adoption of new technologies and ideas throughout human history.

Martin, K., & Quan-Haase, A. (2011). Seeking knowledge: The role of social networks in the adoption of ebooks by historians. In C. Johnson, P. McKenzie & S. Stevenson (Eds.), *Proceedings of the Canadian Association of Information Science.* Fredericton, NB, 1–4 June. Retrieved from www.cais-acsi.ca/proceedings/2011/91_Martin_Quan-Haase.pdf. This paper examines how ebooks are diffusing among scholars of history.

Rich, L. (2010, March 15). Shiny new things. *Ad Age Insights.* Retrieved from http://adage.com/images/bin/pdf/shiny_new_things.pdf. In her essay, Rich examines the influence of early digital adopters in contemporary society and the types of factors they ascribe to their decision-making process in choosing whether or not to adopt a new device.

Rogers, E.M. (2003). *Diffusion of innovations* (5th ed.). New York: Free Press. Rogers' classic work investigates the processes that define and characterize the adoption or rejection of innovations.

Online Resources ....

Apple WWDC 2010 Video
www.apple.com/apple-events/wwdc-2010/
Former Apple CEO Steve Jobs introduces the iPhone 4 at the Worldwide Developers Conference on 7 June 2010, in San Francisco, California.

International Communication Association (ICA)
www–rcf.usc.edu/~ica/
The ICA is an organization linking communications researchers, and includes mass communication as one of its 17 principle divisions. This group also publishes several journals, such as the *Journal of Communication.*

6 The Labour of Technology

Learning Objectives

- ⊛ To examine the role technology has played historically in the division of labour and resulting labour struggles.
- ⊛ To analyze the relationship between technology and immaterial labour.
- ⊛ To critically examine how information and communication technologies (ICTs) are changing organizational structures.
- ⊛ To obtain an overview of the key approaches that are being developed to understand the complexities of digital labour.

Introduction

Technology and labour are linked in many complex ways. For one, there is the labour that results from designing, building, and implementing technology, which we have discussed in Chapter 4. Technology also becomes an extension of our human faculties—both physical and cognitive, allowing us to perform work differently. The goal of this chapter is to elucidate the ways in which technology and work intersect. We discuss changes that have occurred in the nature of work itself as a result of technological advancements, focusing in particular on deskilling, and immaterial and digital labour. We also examine how information and communications technologies (ICTs) are facilitating the development of new organizational structures—often covered under the umbrella term of the networked organization. Finally, the chapter reviews how Web 2.0 technologies facilitate new forms of production based on the principles of collaboration, sharing, and open source. The discussion in this chapter makes it clear that technology is not a neutral mediator in conflicts arising from changes in work conditions. Quite the contrary, technology becomes an active force that changes the nature of work itself, working conditions, and the structure of society as a whole.

Luddites: The Early Struggles with Technology

The application of technology caused a shift in the roles and working conditions of labourers during the Industrial Revolution, particularly in the textile and the agriculture industries. In the textile industry, machines were introduced to simplify and speed up weaving processes. In addition, machinery allowed employers to hire non-skilled workers at lower wages, leading to an increase in unemployment of highly skilled craftspeople (Berg, 1994). To protest against **deskilling**, some workers took action against what they saw as unjust changes and angrily revolted against the machines that they held responsible for the decline in their status and livelihood. The term **machine breaking** describes workers' destruction of technology as a way of demonstrating their frustration with economic, social, and work changes related to industrialization (Hobsbawn, 1952). The act of machine breaking symbolizes workers' perceptions of who was responsible for the changes in work conditions. These acts were centrally coordinated, as the machines were viewed as threats to standards of living, employment, wages, and working conditions (Dinwiddy, 1979; Hobsbawn, 1952).

Key to understanding these social upheavals and the act of machine breaking are the **Luddites**. According to Binfield, "Luddites sought to put an end to the manufacturers' use of certain types of machinery" (2004, p. 3). Specifically, they targeted machines that would lower production costs through decreased wages or work hours. The group reached its political height around 1812, when machinery was becoming more widespread in factories. The Luddites utilized guerrilla tactics, often attacking or even killing the owners of machinery and threatening local officials who collaborated with employers (Binfield, 2004).

The term *Luddites* is connected to the semi-mythical, fictitious leader of the movement named Ned Ludd. According to folklore, **Ned Ludd** was a weaver who was whipped by his master and responded by destroying a knitting frame (Sale, 1995). This rebellious act was viewed as a symbol of the empowering of workers through aggressive behaviour against technology. The name *Ludd* was adopted by followers—mainly dissatisfied workers in the textile industry—who, like Ludd, would engage in acts of machine breaking (Sale, 1995). Ludd's name would often be signed to documents distributed by the Luddites, frequently with the moniker "General Ludd" or the more seditious "King Ludd" attached to the manuscript (Jones, 2006).

Despite all the attention Ned Ludd has received, there is little evidence to suggest he actually existed. His name was likely signed by anonymous Luddites to arouse fear. (The idea of Ludd is also linked to early English folklore of the hero Robin Hood and his Merry Men, who rebelled against the rich and attempted to redistribute wealth to the poor.) Luddism faded in part due to the strict penal consequences given to those engaged in machine breaking, such as deportation to the then penal colony of Australia for a period of 7 to 14 years (Jones, 2006; Sale, 1995). The term **Neo-Luddism** or Luddite continues to be used in the digital era, however, often by advocates of technology to describe in negative terms those individuals who are opposed to or who question technological developments.

A Luddite machine breaker disguised as a woman urges his companions to attack a
factory. Image originally published in May 1812 by Mess, Walker, and Knight.
Source: © Stock Montage/Getty Images.

Technology and the Division of Labour

A central question of this book (see the discussion in Chapter 3) is whether technology is the central actor leading to social change or merely a tool for specific social, political, economic, and cultural purposes. This question also frames our discussion about the intersection of technology and labour. The example about machine breaking shows how even rudimentary equipment could contribute to the deskilling of labour, which then dramatically deteriorated labour conditions for workers in early industrial-era England. We discuss in this section how Taylorism and Fordism are further continuations of this process of deskilling by technology through the division of labour. These two central forms of labour also illustrate how technology ceases to be merely a tool but instead becomes seamlessly integrated into the social and economic system of production.

Taylorism has emerged as a prevalent form of management since the mid-nineteenth century and stems from the work of Frederick Winslow Taylor. Its core principles are based on the idea that work processes can be systematically studied and, based on this knowledge, standardized in such ways that they become more efficient and less prone to error. This approach is also often referred to as **scientific management** because systematic analysis is applied to the study of labour. Taylor developed the technique of **time study**, which consisted of first recording a worker's movements in detail, then breaking the task down into smaller segments, and finally reassembling these segments with the aim of increasing productivity. In this way, management takes over "the burden of gathering together all of the traditional knowledge which in the past has been possessed by the workmen and then of classifying, tabulating, and reducing this knowledge to rules, laws, and formulae" (Braverman, 1974, p. 112). Taylorism radically streamlined work processes, leading to the development of **mass production** techniques by, for example, eliminating wasted time and optimizing workers' motions.

From the standpoint of business owners, Taylorism dramatically helped cut costs. At the same time, it had a number of negative and wide-ranging consequences for work environments. First, the standardization that was involved in streamlining production brought about further deskilling because it made it possible to hire inexpensive, unskilled labour (Braverman, 1974), to which workers reacted with resentment and opposition. Second, instead of workers becoming further specialized in their trade, Taylorism resulted in the division of labour, thereby alienating workers from their craft (Pruijt, 1997). Finally, workers lost control over their work as management imposed standardized production processes. All these factors together led to what Marx has referred to as **alienation** (*Entfremdung* in German), which describes the gap between workers' crafts and their actual work (Ollman, 1976). Workers feel disconnected as tasks become routine, standardized, and ultimately meaningless.

Even though mechanization itself was less of a concern to Taylorism, its principles laid the foundation for the integration of technology in the management of work. Whereas Taylorism required control from management to guarantee that workers performed tasks with maximum efficiency, **Fordism** took the principles outlined in the scientific method a step further by introducing technology as a means to mechanizing, standardizing, and expediting work processes. At the Ford Motor Company, a system was developed where tasks were not only systematically divided to maximize efficiency but were, in addition, mechanically paced through an assembly line, where each worker was responsible for a single, repetitive task. The first successfully mass-produced consumer good was Ford's Model T in 1908, which epitomized the coming of a new social and economic era with the introduction of a **moving assembly line** at the manufacturing facility in Highland Park, Michigan.

The moving assembly line quickly revolutionized production, increasing productivity tenfold and reducing prices dramatically (Hounshell, 1985). Edwards describes how the assembly line "functions as a system of technical control, which means that the entire production process, or large segments of it, are based on a technology which regulates the working pace and controls the labor process" (1979, pp. 112–13).

Not only did technological advancements impact on work processes as they took place at Ford Motor Company and factories around the world, but they also led to major social change. The prices of consumer goods decreased considerably as a result of the mass production of goods, creating a larger consumer base. In addition, central to Ford's economic model was the notion that workers could become consumers of the goods they produced if their wages were increased (Sward, 1948). This led to a cycle of production and consumption that would further feed into the **mass consumption** of goods. Hence, the term *mass production* has come to describe more than just the mechanization of the process of production; it also describes the "interrelated technical, social, and political elements that sustain a unique model of social and economic organization based on the mass consumption of standardized goods" (Lewchuk, 2005).

This section discussed Taylorism and Fordism as two means of production that were made possible through technological advances. Not only did technology replace workers, but when the assembly line is considered, technology dictated the pace and nature of work processes. These two means of production show how in these factories technology becomes more than an external actor; it becomes part of the social and economic system of production. As a continuation of this discussion, the next section demonstrates how technology continued to affect society in the late twentieth century by creating new work structures referred to as networked organizations.

The Networked Organization

Most organizations follow a formal structure or hierarchy to describe work roles and the functions associated with these roles. In these organizational structures, power is distributed from the top down to the bottom levels of the hierarchy. Organizations that follow such a structure are referred to as **hierarchical organizations** and are the most prevalent form of management, in particular among large businesses. These organizations use an **organizational chart** to depict employees' roles that often includes lengthy descriptions of work functions and reporting relationships, indicating who reports to whom. Figure 6.1 depicts the 2009 organizational chart of the **Wikimedia Foundation**, which operates **Wikipedia**, among other wiki projects. At the top of the hierarchy in Figure 6.1 are the executive and deputy directors, who manage and coordinate the technology, the programs, and the finance divisions.

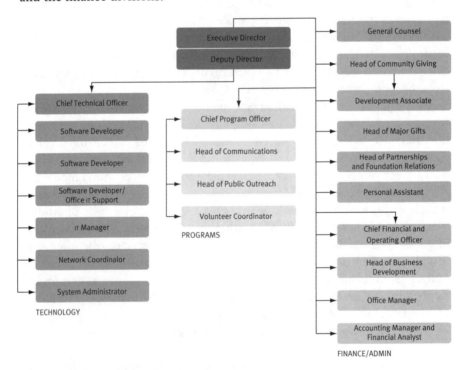

FIGURE 6.1 Wikimedia's Organizational Chart
Source: Wikipedia.org.

The widespread use of the Internet and related information and communication technologies (ICTs) have transformed how people work in organizations. Arguments about the nature of these transformations range from subtle changes in the speed, volume, and ways in which people communicate, to more radical changes consisting of a shift in how power is distributed, how decision making takes place, and how information flows (Bonabeau, 2009; Jarvenpaa & Ives, 1994). The term **networked organization** or network-centric organization is used to describe these new forms of work (Oberg & Walgenbach, 2008; Quan-Haase & Wellman, 2006).

We can observe three major trends in terms of how organizations are embedding technology in their work routines. First, there is a trend toward the paperless office, where information and data are collected, stored, and managed in digital form (Sellen & Harper, 2002). Indeed, many companies outsource data entry to China, India, or Taiwan, where **cybertariats** (Huws, 2003) work day and night shifts inputting information about credit card usage, insurance policies, and medical records. Cybertariats are defined as women, commonly working in developing countries, who are paid minimum wage for routine data-entry jobs. The second trend is the use of mobile technologies to facilitate communication. Smart phones, such as the BlackBerry and iPhone, have blurred the work–home boundary by providing flexibility, mobility, and constant accessibility to employees. The third trend is the use of virtual teams that come together to solve problems on demand. These are teams that are located in different countries or continents that work on a project for specific periods of time, but never meet face to face. Once the project is completed, they are either assigned to a new project or they switch to a different job.

Box 6.1 ✶ kme: A Case Study of a Networked Organization

Here we present Knowledge Management Enterprises (kme) as a case study of a high-tech corporation that has a network structure. Its involvement in knowledge-intensive activity and its high reliance on ICTs make it a good example of a collaborative community in a networked organization. For kme, the exchange of information and the creation of new knowledge are essential, as the firm is under constant competitive pressure to develop and improve its services and products. To remain innovative, kme relies heavily on collaboration among technologically savvy employees using ICTs.

kme shows the following three structural characteristics, which are viewed as central for networked organizations:

1. **Decentralized decision-making.** The company has a vertical form of communication, which enables workers rather than management to be at the centre of information exchanges (Bonabeau, 2009; Castells, 1996; Monge & Contractor, 1997, 2003). Bonabeau suggests that in these new forms of work "many people are empowered to make their own independent decisions"

(Bonabeau, 2009, p. 49) as they can collaborate in real time using various technologies for data gathering, editing, and dissemination.

2. **Local virtualities.** Even though KME employees are collocated, they tend to communicate via digital media, creating dense networks of exchange (Quan-Haase, 2009; Quan-Haase, Cothrel, & Wellman, 2005). The observation that people are no longer interacting visibly in public spaces does not necessarily reflect isolation, but it does show that not all interactions take place in person.

3. **Bridging connections.** At the company, digital means of communication facilitate external communications that bridge organizational boundaries, allowing for the creation of sparsely knit, boundary-spanning structures (Quan-Haase & Wellman, 2008; Sproull & Kiesler, 1991).

Table 6.1 shows the frequency of communication at KME across all media. Digital forms of communication are pervasive even for communicating with those who are nearby. In networked organizations, it is not uncommon to send a text message, instant message, or Facebook wall post to someone who is located in the same room.

TABLE 6.1 **Communication Patterns in a Networked Organization**

Communication at KME (Days Per Year)			
	F2F & Phone	Email	IM
Within department	240	306	306
Elsewhere in organization	99	213	215
Outside organization	21	103	72

F2F = face-to-face

In addition to the changes that have taken place in terms of the use of technology, we can also observe changes in the structure of these organizations. Box 6.1 presents a case study to illustrate the key characteristics of networked organizations and to critically examine the pros and cons of these new work structures.

Even though the concept of the networked organization has become popular, however, the results regarding the effects of ICTs on organizations are inconclusive (Markus & Robey, 1988). Certain kinds of effects, such as speed of communication and increased volume of messages, are apparent within most organizations (Oberg & Walgenbach, 2008; Sproull & Kiesler, 1991).

With regard to changes in organizational structure, the results show that the effect of technology, as we have mentioned before, is not unidirectional and does not occur in predictable ways. ICTs have been found to both centralize and decentralize decision making, and to have no effect when one was expected (Barrett, Grant, & Wailes, 2006; Franz, Roby, & Koeblitz, 1986). For example, in her study of a team of scientists, Roehrs (1998) found that ICTs did not help in making peripheral actors more central in the information network. In this group, technology supported existing work relations and hierarchical structures, but it did not change either in positive or in negative ways how people were connected to each other.

The culture of many organizations, however, is being transformed by new practices that emerge from the digital means of communication. This change is ongoing with the new workforce that consists primarily of **digital natives**, young people who have grown up with the Internet (Palfrey & Gasser, 2008). This generation will not be aware of a pre-Internet time, and ICTs may be normalized in such a way that working in virtual environments may not have the same kind of consequences as it did for previous generations, who only slowly got used to digital media.

In conclusion, the analyses of how ICTs lead to organizational change show that this is a complex social process, where many cultural, organizational, and social factors need consideration. ICTs have not only changed the structure of organizations but have also led to the creation of a new type of worker, one that moves away from industrial-based labour to immaterial labour.

The Immaterial Labour of Technology

Autonomous Marxism, also known as autonomism, is a political movement as well as a theoretical perspective that originated from 1960s Italian Marxist ideology. Antonio Negri, Paolo Virno, and others later developed it into a variant of Marxist social theory with an emphasis on workers' ability to self-organize with the aim of creating changes in the workplace and throughout society at large. The term *autonome* has Greek roots and refers to living within society, following one's own rules.

One of the key thinkers of the autonomous movement was Lazzarato (1996), whose major contribution to social theory was to acknowledge and interpret the changes taking place in certain sectors of the economy. He observed that since the 1960s, new forms of labour were emerging that no longer could be classified as industrial based but, rather, existed around the outputs of the industrial era. These changes in labour resulted from new developments within the production system. With the new systems of production in place, the market was inundated with inexpensive consumer goods. Once most Americans owned a Model T, for example, Ford Motor Company saw demand stall. It was at this point that producers realized that an important part of the production cycle was to create **markets** for their products. For example, General Motors, on the advice of Alfred Sloan, developed five lines of cars—Chevrolet, Pontiac, Oldsmobile, Buick, and Cadillac—that increased in price, reflecting the owners' economic and social status; this was referred to as the *ladder of success*. As a result, an entire industry emerged around marketing, which includes branding a product, developing an advertisement campaign, and finding consumer or niche markets. Lazzarato employed the term **immaterial labour** to describe the industries that support industrial work and encompass "the labor that produces the informational and cultural content of the commodity" (1996, p. 133). Lazzarato (1996) further distinguished between two kinds of immaterial labour.

The first kind of immaterial labour describes the change in the nature of work that was taking place in the 1960s in many corporations. He observed that in addition to the work being done in factories, an increasing number of new workers were doing information-related tasks. The shift consisted of these workers focusing on "selling" rather than "manufacturing" products. The term **white-collar worker** was introduced to describe these professionals or educated workers whose jobs do not entail manual labour. The work they perform is "immediately collective, and we might say that it exists only in the form of networks and flows" (p. 136). And the move away from manufacturing toward a service or information economy is often described as **Post-Fordist** or **post-industrial.**

Bell (1973) characterized this post-industrial society as consisting of three core elements: (1) an increase in services over manufacturing; (2) a dominance of science-based industries; and (3) the creation of a new social class— the **technical or technocratic elite**—who have expertise in the technical areas. Later, with the technology boom of the 1990s, Richard Florida introduced the term *the creative class* to describe workers in the high-tech sector (Florida, 2001; Reese, Faist, & Sands, 2010). He sees these workers as central for economic development of post-industrial cities in North America because they are usually highly educated, creative, open to new ideas, and with higher-than-average incomes.

The second kind of immaterial labour describes the activities and processes that are necessary to shape culture, fashion, tastes, consumer behaviours, and public opinion. Marketing firms, such as the Kaplan Thaler Group, became a part of North American culture through their clever and pervasive branding approach. For example, in 1999 only 10 per cent of Americans were aware of the Aflac brand, an insurance company. With a series of commercials featuring the "Aflac Duck," the Kaplan Thaler Group increased Aflac's brand awareness to 94 per cent and increased sales by 55 per cent. Now the Aflac Duck has become a part of American popular culture. The marketing strategy that the Kaplan Thaler Group used is comprehensive and includes, in addition to traditional television and radio advertisements, social media presence via a Facebook page, a Twitter account, and YouTube videos, demonstrating a huge change in the strategy used to sell products to consumers.

While technology radically transformed the nature and pace of work during the industrial era, in the post-industrial society technology created a radically new form of work. Immaterial labour directly exists in relation to technology, which helps in creating, transmitting, repackaging, and diffusing information about products on a global scale (Dyer-Witheford, 2001). The aim is no longer to produce new merchandise but to create a savvy marketing strategy aimed at promoting and selling products. In the next section, we will discuss the value of user-generated content and resulting unpaid immaterial labour as it takes place in Web 2.0 environments.

The Value of User-Generated Content and Unwaged Immaterial Labour

Immaterial labour is also reflected in the type of work that users in **Web 2.0** environments perform. Web 2.0 developed as an alternative form of communicating on the Web; instead of information being static, Web 2.0 offers a range of features that facilitate among other things interactivity, collaboration, and feedback.

At the centre of Web 2.0 is the concept of **user-generated content**, which characterizes a shift from content being centrally produced and distributed to a model where it is created, produced, and edited by **end-users**. The value of user-generated content is in most cases not associated with a price tag because its produsers do not obtain remuneration. In contrast to immaterial labour that takes place in the context of paid work, users of social media, such as Flickr or Twitter, do not get paid. They are motivated to participate because they are a part of a shared, collaborative community (Bell, 1973; Bruns, 2008). The affective relationships that are formed in the context of the community are central for continued participation. There are also personal gains to be made through participation. Members of the community can become central figures with strong reputations for their expertise through such functions as code development, community coordination, or administrative service.

Box 6.2 discusses Facebook as an example of a service that relies on Web 2.0 immaterial labour.

Box 6.2 ❖ FACEBOOK AS WEB 2.0 IMMATERIAL LABOUR

Facebook is a good example of a Web 2.0 application that is based on immaterial labour. In 2010, there were 500 million worldwide users of Facebook, of which Canada had the highest penetration rate (see Figure 6.2). Young Canadians use Facebook more than any other communication technology, including email, texting, and cellphone, and have an average of 190 friends in comparison to 130 for the rest of the world (Moretti, 2010). Facebook as a corporation provides the infrastructure for users to add personal information. In turn, users generate content daily for other users by adding pictures, writing on friends' walls, adding comments to posts, rating posts, and linking to internal and external information. While most users would not consider this work but, rather, art, pleasure, fun, and leisure time, this investment of time and content constitutes revenue for Facebook. Facebook is valued at about US$100 billion—a disproportionate amount for a social networking tool that facilitates social exchanges. Its value becomes quickly apparent when we examine the capital investment so far put into Facebook. For example, Microsoft invested a total of US$240 million in Facebook for a slim 1.6 per cent stake (Stone, 2007).

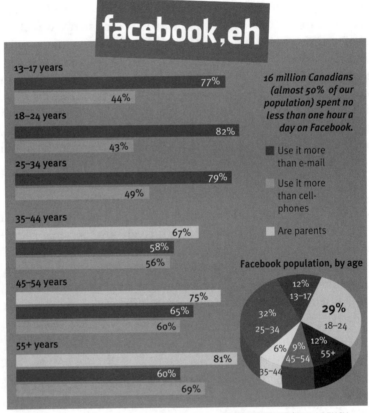

MEGAN DINNER/QMI AGENCY

Figure 6.2 Facebook's User Growth in Canada
Source: Moretti, 2010.

What are the assets that give Facebook such high value? The most valued asset is the revenues gained through direct advertisement to users. Users are exposed to a series of advertisements by simply logging on to the service, often tagged as "sponsored." A recent study found that Canadians are interacting frequently with brands. Of the estimated 15 million Canadians on Facebook in 2010, nearly 1.2 million are fans of the Tim Hortons Facebook profile (Moretti, 2010).[1] Interestingly, the number of hits to the Starbucks website is much lower in comparison to the hits to its Facebook page: 2.7 million and 16.3 million a month, respectively. This represents a drastic gap and shows the increasing relevance of social media as a means to reach out to consumers.

In addition, Sheryl Sandberg (as quoted in Moretti, 2010), who is Facebook's chief operating officer, asserted that "[a] recommendation by a friend is the best kind of marketing," showing the power of peer-to-peer recommendations. The third value of Facebook is to be able to understand users and consumers through their interactions with the content. Facebook provides data about users' opinions about content and their likes and dislikes of advertisements, companies, and products. This is very much the kind of information that would be difficult for marketers to obtain from such large numbers of individuals. About 50 per cent of Canadians spent no less than an hour on Facebook every day updating their profile, posting on friends' walls, and adding content. To some extent, Facebook users are "working" while on social network sites by providing important information for marketers about product trends, consumer preferences and opinions, and the value of products.

The labour that takes place in Web 2.0 environments is unique because it is unpaid. People who actively participate in these environments contribute because they feel personally involved. As boyd and Heer (2006) have argued, people write themselves into being on social media and express through digital traces their identity, their social webs, and their daily activities. Profiles are becoming extensions of who they are (Hogan & Quan-Haase, 2010). In this context, users' expression of self becomes closely interlinked, or almost inseparable, from their digital labour.

The Role of the Prosumer

ICTs have not only changed the nature of labour, but have also profoundly modified the relationship between the consumer, the producer, and the product. The first to acknowledge this change was Alvin Toffler (1980), who introduced the concept of **prosumer** or produser in the 1980s to describe the merging of producer and consumer. Consumers become active in the production process by performing tasks that used to be delegated to producers. Many do-it-yourself (DIY) kits are available that allow consumers to be empowered and independent from producers. For example, the IKEA model of putting furniture together oneself without the help of a service person, sometimes accompanied with high levels of frustration and low levels of success, puts this process into the hands of consumers.

Following Toffler, in 2006 Don Tapscott and Anthony Williams introduced the idea of **prosumption** by combining the words *production* and *consumption*. *Prosumption* describes new modes of production where customers play a central role in the design, development, and use of the end product (Tapscott & Williams, 2006). An example is *Second Life*, a virtual world in which its users are given a great degree of freedom to shape the product by creating their own avatars and building digital artifacts in a 3-D environment. In prosumption business models, corporations will often provide only the infrastructure, and prosumers will be in charge of producing content. In these cases, prosumers are not only co-producers but often also drive the creation of the product because there would be no product without their active participation.

Produsage is a term coined by Axel Bruns (2008) and is a modification to the concept of prosumption. Bruns uses the term *produsage* to describe the shift in the mode of production that is characteristic of the labour taking place within Web 2.0 environments. This new model blurs the line between producers and consumers and empowers users as they take control over the production and the distribution of content. Bruns describes producers as those who

> participate in the development of open source software, in the collaborative extension and editing of the *Wikipedia*, in the communal world-building of *Second Life*, or processes of massively parallelized and decentralized creativity and innovation in myriads of enthusiast communities do no longer produce content, ideas, and knowledge in a way that resembles traditional, industrial modes of production; the outcomes of their work similarly retain only few of the features of conventional products, even though frequently they are able to substitute for the outputs of commercial production processes. User-led "production" is instead built on iterative, evolutionary development models in which often very large communities of participants make a number of usually very small, incremental changes to the established knowledge base, thereby enabling a gradual improvement in quality which—under the right conditions—can nonetheless outpace the speed of production development in the conventional, industrial model. (2008, p. 1)

In the produsage model, the end goal is the creation of an **information commons** (Hippel, 2005), which provides the social structure for supporting the core activities of the community, such as sharing, collaborating, negotiating, and developing. This information commons—an amalgamation of nodes consisting of content, users, and producers—then ultimately allows the information to continue to be relevant and up-to-date in the context of the community. Any information system in which people actively collaborate to generate content can be considered an information commons, including Wikipedia, Facebook, and Flickr.

Most of the labour performed by produsage communities is offered for free, without expectation of a wage, benefits, or any other kind of return. This kind of work differs fundamentally from the kind of work done by waged immaterial labourers in corporate environments, as discussed in the section earlier in the chapter. Involvement in produsage communities is motivated by an interest in the content itself, by the affective relations that develop among participants in a produsage project, and by the satisfaction of creating a product that will serve the community.

There are exceptions, however, as with Amazon's Mechanical Turk (see www.mturk.com/mturk/welcome), which pays participants for performing small problem-oriented tasks. Here, a global workforce comes together to perform any digital task necessary, from responding to surveys, to analyzing marketing strategies, to providing suggestions for slogans. Mechanical Turk pays its employees per task and facilitates access to a global, diverse, and on-demand workforce that has no contracts, no benefits, no job security, no boss, and no structure.

Preconditions of Produsage

Central to our understanding of collective action and hence collaborative content production is the coming together of technical possibilities, social organization, and collective features. To understand the processes better, Bruns (2008) has created a model of produsage by describing four preconditions and four key principles of distributed and collaborative content production. We will start by introducing the four preconditions because they represent the factors that are needed for produsage to emerge, and then we will discuss the core of the four principles. For Bruns, each of the four preconditions shapes how content can be produced in a collective manner.

1. *Probabilistic, not directed problem-solving.* In bureaucratic forms of organization, problems are solved from the top to the bottom, with management giving direction as to who will take over the task and how. By contrast, in produsage projects there is no centralized decision making because all members are volunteers. This leads to what Bauwens (2005) has described as **holoptism**, an open concept to problem solving, where all members of the group have equal insight into all components of the

problem. A problem gets solved as members of the community identify it as relevant and work relentlessly to find a solution.

2. *Equipotentiality, not hierarchy.* Produsage projects are based on the notion of **equipotentiality**, which refers to the ability and possibility of all members contributing equally to the end result. Participation is not limited by hierarchical roles but is instead all-inclusive. As well, participation in projects is flexible and not based on authority; participants can bring their expertise to bear on any aspect of the project where they feel they can make a contribution instead of being confined to work on pre-assigned elements.

3. *Granular, not composite tasks.* This precondition is based on the work of Benkler (2006), who argues that "the number of people who can, in principle, participate in a project is . . . inversely related to the size of the smallest-scale contribution necessary to produce a usable module" (p. 200). With projects that do not require much administration, that is, where the tasks are independent, it is possible to complete the project quickly when many people can participate simultaneously.

4. *Shared, not owned content.* The fact that the content is shared among project members is one of the primary distinctions between produsage and industrial, Fordist-based models of production. Industrial models rely on the notions of ownership and secrecy to impart information, which is only available through the formal channels of command on a need-to-know, top-down basis. By contrast, produsage models of production are more effective because information flows freely and is shared, distributed, and re-used in a fair manner among community members.

Key Principles of Produsage

Once the preconditions are in place, open, participatory communities can form to produce content collaboratively. Bruns (2008) proposes four key principles that are characteristic of different produsage projects. Not all principles are present in all produsage projects, but at least some of them need to be in place for such projects to be successful. Some of the key elements highlighted are as follows.

1. Open Participation, Communal Evaluation

Quality control occurs very differently in a produsage model than it does in a hierarchical form of organization. When an idea or a contribution is proposed, the community of produsers evaluates it, either by embracing it and further developing it or by ignoring it. The core idea in produsage is that when many participants provide feedback on a contribution, this will lead toward a better outcome. From this viewpoint, in a produsage community all contributions, regardless of how small they may be, are relevant to the quality and relevance of the final outcome. Participants from a wide range of backgrounds, talents, and levels of expertise are encouraged to participate, making these communities highly inclusive and heterogeneous.

2. Fluid Heterarchy, Ad Hoc Meritocracy

Instead of functioning as hierarchical institutions, produsage communities are based on input and expertise. This makes them fluid because individuals can contribute to a specific problem or help with a specific task, but then, when the task is completed, they become less active and move on to other projects. Another important aspect of this form of organization is that it allows participants to individually or in small teams make decisions about aspects of the project without requiring consultation or top-down approval, making produsage projects more effective and problem oriented.

The notion of fluid heterarchy has its origins in the writings of the economist Adam Smith (Smith, Campbell, & Skinner, 1981). Smith saw the division of labour as one of the main improvements in productivity because it takes advantage of specialized skills and expertise. By dividing the overall production process into smaller tasks, expertise can be applied more strategically and the efficiency of the entire process can be increased. Produsage projects work as loosely connected networks that do not follow a strict hierarchy of command and control but instead consist of several teams working simultaneously on various components of the problem.

The concept of ad hoc meritocracy comes from the work of Alvin Toffler (1970), who described the coming of **adhocracy** as a new form of organization. In adhocracy, individuals or teams are assembled as they are needed to solve narrowly defined, short-term problems instead of having permanently assigned roles and functions based on organizational charts as described in the earlier section on the networked organization. In the context of produsage, these ad hoc teams are also based on the reputation that participants build as leaders and as experts in the community; hence, participation and relevance are based on merit.

3. Unfinished Artifacts, Continuing Process

The work structure of a produsage project, with its fluid and equipotential participation, its granular approach to the division of labour, and its network-centric shape, naturally leads to projects that do not yield finished end products. Rather, projects and artifacts remain works-in-progress with participants continuing to add tasks and to improve existing content. Bruns (2008) explains how "produsage does not work towards the completion of products (for distribution to end users or consumers); instead, it is engaged in an iterative, evolutionary process aimed at the gradual improvement of the community's shared content" (p. 27).

What distinguishes outcomes originating from produsage projects is that they become cultural artifacts instead of commercial products. When we look at produsage projects, we see a snapshot of the process, but the product will continue to be modified as users add, re-examine, or modify content, or even make changes to the interface itself. That is, the entire product of a produsage project is in a constant state of flux, including the parameters around content production.

The extreme case of unfinished artifacts is represented in software that is referred to as **perpetual beta**. This is software released to users under the understanding that it is incomplete and in a state of improvement. The main rationale for releasing software before its completion is to receive feedback from users as the product is being developed and to alter the website or service to reflect some of the suggestions made by users.[2] Figure 6.3 shows the key characteristics of projects that work with perpetual beta releases. O'Reilly (2005), who coined the term *Web 2.0*, argues that it also makes sense to continually release new versions and develop the software in the open because software is in constant change; as a result, setting a release deadline only compromises the product's quality.

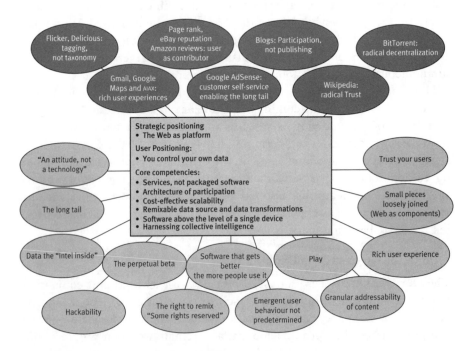

FIGURE 6.3 Characteristics of Perpetual Beta
Source: Tim O'Reilly, 'What Is Web 2.0?', http://oreilly.com/pub/a/web2/archive/what-is-web-20.html?page=1. © 2005 O'Reilly Media, Inc. All rights reserved. Used with Permission.

A number of companies use this model, where they work closely with the end-user and release various beta versions as development continues. For example, Gmail, Flickr, and del.icio.us often feature a "Beta" logo to indicate their ongoing development of the product (O'Reilly, 2005) instead of working with major releases as Microsoft continues to do with its release of Windows XP, Vista, Windows 7, and the next generation of Windows products.

4. Common Property, Individual Rewards

At the heart of produsage is the idea that content should be available to all users, even those who may not have directly contributed to the project. Users should be able to not only view the content but also to use the content creatively for the purpose of creating new content, ideas, or products. To guarantee that intellectual property created within and by the community is properly protected, a set of standards around copyright and legitimate uses of content have been generated. Some of the basic rules surrounding the use of content in information commons include the following (see Bruns, 2008):

- Content created in the community must be available at no cost.
- If the content is modified, the newly created content must be made available to the community under comparable copyright agreements.
- The creators of the content must be properly acknowledged and, if necessary, remunerated.

A number of different websites provide information about the available legal documents to protect copyright and the provisions articulated in them. The most well-known are the GNU **General Public License**, the GNU Free Documentation License, the Open Source License, and the Creative Commons license framework. All of these documents are legally binding.

In sum, Bruns (2008) proposed the four preconditions and the four key principles to produsage as a means to analyze work and work structures as they take place in digital environments. While produsage communities do not follow strict hierarchies, they still require some form of regulation to operate. Regulations are particularly important to solve disputes among contributors. For example, in Wikipedia it is common for individuals or groups to emerge with divergent views on sensitive or controversial topics. When Michael Jackson died on 25 June 2009, his Wikipedia page was being constantly updated (about 100–200 changes per day). A battle emerged between those who revered him and wanted to portray him in the best possible light and those who were critical of him and wanted to emphasize the accusations of child molestation.

For the community to function, a comprehensive list of rules has been developed that users need to follow if they do not want to be banned from participation. There are a number of ways in which disputes among participants can be solved. For example, participants can discuss their differences "through comments and annotations 'in the margins,' and by the repeated overwriting of existing passages in a shared effort to arrive at a better representation of communally held values and ideas" (Bruns, 2008, p. 27). As well, people may look at the history of individual entries and participants' past contributions to determine the evolution of the content. The most fundamental principle guiding participation in Wikipedia is the **neutral point of view** (NPOV), which expects participants to represent in each article "fairly, proportionately, and as far as possible without bias, all significant views that have been published by reliable sources" (Wikipedia, 2011). Hence, a combination of community- and technology-based processes provides the necessary context for produsage communities to function.

Conclusions

The chapter has shown that the relationship between technology and work is not at all straightforward. The introduction of mechanized technologies during the Industrial Revolution alienated labourers from their work. Prior to industrialization, tools had been a part of craftsmanship, and had been used as extensions of the human body. Industrialization brought about radical change, by directly substituting human labour with machinery, as well as skilled workers with cheap labour. With Taylorism, the process of deskilling continued with further standardization and division of labour. Fordism not only took deskilling to the next level but introduced technology as a means of controlling work processes. Under these conditions, workers no longer had agency over their work; instead, they became a cog in the wheel.

Yet in post-industrial societies, technology is no longer a substitute for the worker but, rather, a vehicle for individuals to apply information-based skills in order to develop, produce, and enhance their labour. The advent of immaterial labour represented a turning point, where instead of further deskilling workers, technology created labour around the creation, transmission, diffusion, and repackaging of information.

In Web 2.0 environments, we see new forms of digital labour emerge through participation in produsage projects. These new forms of work are in stark opposition to hierarchical forms of organizations in that they emphasize collaboration, sharing of resources, and increased communication. Technological advances have reshaped the nature of work as well as the conditions and relations within these environments. Technology emerges as an agent of change, both positive and negative.

The introduction of technologies into the workplace has eliminated traditional professions, yet has also emerged as the cornerstone of new vocations. Information technologies in particular can also become an integral component in modern collaborative works, which connect a new generation of globalized workers and information-hungry consumers in order that they may embark on new projects and situations. Indeed, within our twenty-first-century information society, the role of community has taken on an increasingly participative function in determining the effectiveness and applicability of technology. As we shall discuss in later chapters, the conditions upon which the success of a technology rests often reside within the socio-economic and cultural dynamics inherent within a community.

Questions for Critical Thought

1. Explain the concept of Neo-Luddism and trace its roots to the Luddite movement in Industrial Revolution–era England.

2. Using Flickr as an example, discuss how the concept of immaterial labour helps explain Web 2.0 applications.

3. Explain how the concept of the networked organization contrasts with hierarchical forms of work. Do you think that some level of organizational structure is still needed in organizations for the purpose of coordination? Explain.

4. Is it possible to manage and control the immaterial labour performed by workers in the twenty-first century in a similar manner as the labour performed by industrial workers in the nineteenth and twentieth centuries?

Suggested Readings

Brown, B., & Quan-Haase, A. (forthcoming). 'A workers' inquiry 2.0': An ethnographic method for the study of produsage in social media contexts. *TripleC, 10*(2). www.triple-c.se/

Bruns, A. (2008). *Blogs, Wikipedia, Second Life, and beyond: From production to produsage.* New York: Peter Lang. The author discusses the term produsage in great detail and presents current practices of produsage.

Dyer-Witheford, N. (2001). Empire, immaterial labor, the new combinations, and the global worker. *Rethinking Marxism, 13*(3/4): 70–80. The article addresses the complexity of immaterial labour and its relation to the global worker.

Fuchs, C. (2008). *Internet and society: Social theory in the information age.* New York: Routledge. The book connects critical theory to the study of the Internet.

Shirky, C. (2008). *Here comes everybody: The power of organizing without organizations.* New York: Penguin Press. This book shows how the Internet has changed the publishing industry by allowing consumers to be active producers, in a similar manner as the printing press did in the fifteenth century.

Tappscott, D. & Williams, A.D. (2006). *Wikinomics: How mass collaboration changes everything*. New York: Portfolio. Explains how new forms of virtual collaboration are a powerful means to work together.

Online Resources

The Fibreculture Journal
http://fibreculturejournal.org/
   This journal publishes cutting-edge research on digital media, networks, and transdisciplinary critique.

The Foundation for P2P Alternatives
http://p2pfoundation.net/
   The mandate of the foundation is to obtain a better understanding of the social impact of peer-to-peer technology and thought.

Mashable
http://mashable.com/
   The site provides up-to-date information on the technology industry.

Mute Magazine
www.metamute.org/
   The key topics of Mute Magazine are culture, arts, and politics as they take place online.

Produsage
http://produsage.org/
   This resource is maintained by Bruns and provides a general overview of the theory of produsage.

7 Technology and Inequality

Learning Objectives

- ⊕ To discuss the intricate relation between inequality and technology.
- ⊕ To see the evolution of the digital divide from a problem solely of access to a complex phenomenon relevant to all members of society.
- ⊕ To understand the global digital divide and the micro- and macro-level barriers to access and use that exist in developing nations.
- ⊕ To understand the term *technology transfer* as it applies to developing nations.
- ⊕ To examine critical perspectives of the digital divide.

Introduction

In this chapter, we address issues of inequality as they play out in technological use, implementation, and impact. Understanding technological inequality is of great importance to researchers, policy-makers, and politicians because technology provides educational, political, and economic advantages, creating power imbalances and, potentially, conflict between social groups. The chapter specifically examines digital inequality as a pressing issue of our times: what are the social, economic, and cultural consequences for those disconnected from the Internet?

While researchers agree on the importance of studying the digital divide, much controversy still surrounds its definition (Epstein, Nisbet, & Gillespie, 2011; Stevenson, 2009; Vehovar, Sicherl, Husing, & Donicar, 2009). This chapter covers the historical developments of the digital divide concept, examines the complexity of its measurement, and considers its relevance to policy in Canada and the United States. A key argument is that the use of digital technologies not only reflects access but also reveals differences in skill level. Many individuals, despite having access, cannot take full advantage of the resources available to them online because they do not know how to navigate and evaluate the endless number of resources available.

We also cover the term *global digital divide* to describe the gap in access to and use of the Internet that exists between **developing** and **developed nations**. Many developing nations, such as Somalia, continue to struggle in their effort to become digital, having to overcome numerous barriers of access, skill, and infrastructure. Nonetheless, penetration rates have increased in these nations; surprisingly, individuals often gain access to the Internet not through computers but through cellphones. We discuss China as a prime example of a **newly industrializing nation** that has struggled to join the **information society** and in the process has developed an ambivalent relationship with the Internet. The Chinese example shows how the impact of the Internet on society is complex and cannot be limited to a single sphere. The chapter ends with a critical reflection and a series of conclusions about our current state of knowledge regarding the Internet's role in creating social change.

The Digital Divide

The term **digital divide** describes discrepancies between social groups in access to, use of, and empowerment by networked computers and other digital tools. Initially, policy-makers and scholars used the term to describe the gap in personal computer ownership that existed between various social groups. However, with the rapid expansion in the networking of computers in the 1990s, *digital divide* came to denote access to digital resources primarily via computers but possibly also via cellphones, laptops, and e-book readers (Epstein et al., 2011; Stevenson, 2009; Vehovar et al., 2009). The term gained particular prominence when, in a 1996 speech, US president Bill Clinton and US vice-president Al Gore used the term to emphasize the social and technological challenges presented to America in the **information age**, where there would be an increasing gap between **haves** and **have-nots**. Closing the divide turned into a key political target within the North American policy framework (Epstein et al., 2011).

Issues of the digital divide have been phrased in terms of how the Internet provides the opportunities for overcoming or exacerbating existing inequalities. Norris writes about how "digital networks have the potential to broaden and enhance access to information and communications for remote rural areas and poorer neighbourhoods, to strengthen the process of democratization under transitional regimes, and to ameliorate the endemic problems of poverty in the developing world" (2001, p. 6). Consequently, questions regarding the digital divide are of major concern to all citizens because the economic, cultural, and social possibilities of individuals and nations can depend on people's ability to leverage digital technologies and participate in the information age. Also important to consider and critically assess are the early studies of the digital divide and their focus on access. These studies show how the debate around the digital divide has changed over time as well as the kind of challenges early users of the Web needed to overcome.

Early Studies of the Digital Divide

Early studies of the digital divide were primarily concerned with issues surrounding access. **Access** is defined as having a computer at home that is connected to the Internet. Key to examining access has been an assessment of the Internet's **penetration rate** in various social groups. Two US surveys conducted in 1995 and 1998 by the **National Telecommunications and Information Administration** (NTIA)—*Falling through the Net: A Survey of the "Have Nots" in Rural and Urban America* and *Falling through the Net: Defining the Digital Divide*—pointed out alarming differences in Internet penetration rates based on age, gender, education, rural/urban regions, single/dual parent households, ethnicity, and income (Epstein et al., 2011). The results show that inequalities existent in society are reflected in Internet use with users being primarily white, younger males with higher socioeconomic status living in urban areas.

The US debate was mirrored in Canada, where a number of surveys were conducted in the mid-1990s to assess the extent to which Canadians were connected. A Statistics Canada report entitled *Access to the Information Highway* investigated in more detail the effect of income on connectivity (Dickinson & Sciadas, 1996). A comparison of 1994 and 1995 household penetration rates for various ICTs shows two trends: (1) telephone and cable both have reached high penetration rates and these are stable; and (2) a dramatic increase can be observed in the availability of computers and modems. In 1994, 99 per cent of Canadian households had a telephone, 74 per cent had cable, 25 per cent had a computer, and only 8 per cent had a modem.[1] By contrast, in the 1995 data 99 per cent had telephones, 73 per cent had cable, 29 per cent owned a computer, and 12 per cent had a modem. While there is a demonstrated increase in computer and modem ownership within the one-year period, it remains unclear how the technologies are diffusing in society.

For this purpose, the penetration rates were calculated separately for each **income quartile**. Table 7.1 shows that for telephone and cable penetration rates little to no differences can be observed between the high and low income quartiles (as shown by statistical analysis). By contrast, income is strongly associated with both computer and modem penetration rates. In the highest income quartile—those who earn more than C$63,034 a year—computer and modem ownership was 50 per cent and 22 per cent, respectively. Individuals in the lowest income quartile—those who make less than C$21,398 a year—report much lower percentages of computer and modem ownership: 12 per cent and 5 per cent, respectively. Dickinson and Sciadas raise the issue of inequality: "As penetration rates rise, households that can least afford their own terminals will make up a larger proportion of the 'have-nots'" (1996, p. 6). We conclude that in the 1990s affordability was one of the main constraints to household ownership of a computer and modem.

TABLE 7.1 Penetration Rates and Income, 1995

	Income quartiles				
	Bottom	Second	Third	Top	All
	←$21,398	$21,398–$39,949	$39,950–$63,034	→$63,034	%
Telephone	96.0	98.8	99.5	99.7	98.5
Cable	64.4	70.3	76.7	82.2	73.4
Computer	12.3	20.2	32.5	50.2	28.8
Modem*	4.8	7.3	13.6	22.4	12.0

* Average annual rates compounded

Source: Dickinson, P., & Sciadas, G. (1996). *Access to the Information Highway*. Statistics Canada Report. Retrieved from www.statcan.gc.ca/pub/63f0002x/63f0002x1996009-eng.pdf. Reproduced and distributed on an "as is" basis with the permission of Statistics Canada.

In the Statistics Canada report (1996), similar gaps were found along other key socio-economic variables. For example, when the head of the household was employed, computer ownership was 38 per cent, while only 21 per cent and 13 per cent of those unemployed or out of the labour force, respectively, reported owning a computer. Households where the head had a university degree were six times more likely to own a computer than households were the head had less than a grade 9 education. Stark differences in penetration rate are also observed between rural and urban areas: 22 per cent and 6 per cent of rural residents owned a computer and modem, respectively, while 30 per cent and 13 per cent of urban residents owned a computer and modem, respectively. The data suggest that in Canada employment, education, and rural/urban location are all central predictors of computer and modem ownership, creating a major challenge for the government to overcome the disparities in access.

What conclusions can we draw based on the early analyses of Internet penetration rates in Canada and the United States? First, access to the Internet reflects existing inequalities in society with income, employment, education, ethnicity, rural/urban location, and age all affecting adoption patterns. This is not surprising as economic resources are required to purchase a computer and to connect to the Web via an Internet service provider (ISP). Also, education and age play a role because users need skills to utilize a computer and navigate the Internet. This was particularly the case in the early 1990s when graphical Web browsers did not yet exist to facilitate the navigation and retrieval of content. Second, the gap in access may exacerbate existing inequalities by putting the have-nots at a disadvantage over the haves in terms of information, connectivity, and skills. Since these early studies of the digital divide, a number of changes have occurred. Of particular relevance have been initiatives to decrease the digital divide.

Policy Intervention vis-à-vis the Digital Divide

The large discrepancies in access to the Internet brought the issue of the digital divide into public consciousness and triggered new policies targeted at increasing digital inclusion both in the United States as well as in Canada. What kinds of interventions were devised to address issues of the digital divide? Programs were created not only by governments but in parallel by non-profit organizations and private funding agencies. In this section, three of the most well-known interventions will be covered: CANARIE, E-Rate, and the Bill and Melinda Gates Foundation.

The Canadian Advanced Network and Research for Industry and Education, or CANARIE, is a Canadian non-profit corporation backed by the Canadian government. Founded in 1993 and based in Ottawa, CANARIE is a fibre-optic and satellite data network that "facilitates leading edge research and big science across Canada and around the world" (CANARIE, 2010). CANARIE's networking capabilities allow researchers to collaboratively analyze data that can lead to discoveries in the realms of education, health sciences, and technology.

In the United States, **E-Rate** was established in 1997 to provide schools and libraries affordable access to the Internet. E-Rate is the abbreviation for the Schools and Libraries Program of the Universal Service Fund, which is administered under the direction of the Federal Communications Commission (FCC). E-Rate is built on the concept that "technology has great power to enhance education" (Federal Communications Commission [FCC], 2004).[2] The program is chiefly designed to assist American schools and libraries in acquiring telecommunications equipment and services. E-Rate has been effective in providing Internet access in public classrooms and in libraries. Indeed, the number of public classrooms with Internet access increased from 14 per cent in 1996 to 95 per cent in 2005 (Education and Library Networks Coalition, 2007). E-Rate provides discounts to schools and libraries of between 20 and 90 per cent, depending on poverty levels. However, E-Rate does not cover expenditures such as hardware, software, or staff training (Puma, Chaplin, & Pape, 2000). The program has spent a total of US $19 billion on discount services since its inception.

The Bill and Melinda Gates Foundation, founded in 1999, is the largest private foundation in the world (Beckett, 2010). Built upon 15 guiding principles, the foundation embodies a belief that science and technology can perform an important function in improving the lives of those who are less fortunate (Bill and Melinda Gates Foundation, 2010). The foundation's central concerns are in improvements in areas such as access to health care, education, agriculture, and technology for the poor and impoverished. Outside of the United States, the regions the foundation devotes most resources to are Africa and southern Asia (Beckett, 2010).

The three examples discussed above—CANARIE, E-Rate, and the Bill and Melinda Gates Foundation—show the range of initiatives developed with the aim of closing the digital divide. Despite the good intentions of these initiatives, however, the digital divide is a complex problem that cannot easily be solved with the provision of infrastructure, computing resources, and funds. In the next section we examine the effectiveness of the initiatives and their potential to effect real social change.

Closing the Digital Divide

The previous sections discussed the existing gap between the haves and have-nots as well as the various policy strategies in place. How effective have these strategies been in creating real social change? Have penetration rates increased in Canada and the United States along the lines of inequality?

Data from the 2009 *Canadian Internet Use Survey* show that Internet penetration continues to grow steadily in Canada. The survey found that 80 per cent of Canadians aged 16 and older indicated having been online in the 12 months prior to the survey (Statistics Canada, 2009). This contrasts with 73 per cent in 2007, 68 per cent in 2005, 12 per cent in 1995 and 8 per cent in 1994 (The Canadian Press, 2008). While these penetration rates suggest that more Canadians are online than ever before, they do not specifically address the issue of the digital divide. In order to assess the impact of policy initiatives, we must examine penetration rates along socio-economic factors.

More detailed analyses using complex analytical techniques, such as multiple regression analysis, are unfortunately often lacking. Statistics Canada conducted analysis of 2009 data from the *Canadian Internet Use Survey*, where it compared penetration rates by income and rural/urban dimensions (Statistics Canada, 2009). Results show that 9 out of every 10 individuals in the $85,000 per year income bracket use the Internet regularly, in comparison to only 5 out of every 10 individuals in the less than $30,000 per year bracket. When examining education, the report found that in 2009 89 per cent of individuals with at least some post-secondary education used the Internet, compared with 66 per cent among those with no post-secondary education. These more detailed analyses reveal that the increase in penetration rates comes primarily from middle-class users as well as younger generations. Hence, the divide has not been completely closed: the gap between haves and have-nots continues to exist along the lines of income, education, employment, age, and rural/urban location. There are other key aspects that are often disregarded in the debate in addition to access. In the next section, we examine some of these factors, including types of access, knowledge levels, and skills.

Looking Beyond the Access Divide

While the gap in Internet access has shrunk considerably since the mid-1990s, scholars have identified variations among users in terms of the types of access they have to the Web (dial-up, broadband, and wireless), their knowledge and skills when online, and the variation in online activities they perform (Vehovar et al., 2009). Mossberger, Tolbert, and Stansbury write "[i]f some individuals cannot use computer technology, then all the access in the world will do no good. Further, if people cannot find the assistance they need to use the technology, then access alone does little to alleviate the problem" (2003, p. 39). Hence, access to the Internet may be less of a concern over time in Canada and the United States, but differences among Internet users in their skill level continue to be a concern. As a result, alternative definitions of the digital divide have been proposed that better reflect the complexity of the concept. These definitions look at the digital divide beyond the access divide alone and examine how people use ICTs and the kinds of benefits they obtain from this use.

One alternative definition is the one proposed by Mossberger, Tolbert, and Stansbury (2003), where the authors identify four central components of the digital divide: (1) the access divide, (2) the skills divide, (3) the economic opportunity divide, and (4) the democratic divide. We discuss each dimension next and how it affects individuals' lives and their opportunity to leverage resources.

The Access Divide

The access divide examines whether or not a person has access to the Internet and the type of access as well as the location and frequency of use. In the early days of the Internet, having a dial-up connection provided access to a wealth of information. This is no longer the case because the Internet has evolved from a primarily text-based information repository to a multimedia immersive environment. Dial-up does not adequately support Web pages with images, video, and/or voice, making it excruciatingly difficult and frustrating for people with dial-up access to take full advantage of the Web. According to a 2008 Pew Internet & American Life report, 55 per cent of Americans have broadband Internet connection at home. This figure has dramatically increased since 2000 and 2005 when broadband connection was less than 5 per cent and 30 per cent, respectively.

Still, as much as 10 per cent of users continue to have dial-up Internet connection (Horrigan, 2008), reflecting a continued divide along access type by those who cannot afford it. When asked why they did not give up their dial-up connection, 35 per cent reported that price was the key reason for continued use, 19 per cent indicated that nothing would make them switch to broadband, and 10 per cent said that broadband was not available where they lived. Hence, affordability and the rural/urban divide continue to be barriers to equal access to broadband in the United States. Furthermore, the percentage of Americans with broadband Internet access increased to nearly 70 per cent by October 2010 according to the National Telecommunications and Information Administration (2011). This suggests that while the digital divide persists in the United States access to the Internet is narrowing.

The Skills Divide

The **skills divide** can be examined in terms of **technical competence** and **information literacy**. Technical competence includes knowledge about how to use computers, whereas information literacy refers to the ability to seek out information, evaluate it, and use it for specific purposes, such as finding a job, solving a problem, or dealing with health-related concerns.

In Canada and the United States, attempts to close skills-based gaps in the digital divide occurred early on through the establishment of community-based training programs. In Canada, the **Community Access Program (CAP)**, a government initiative administered through Industry Canada beginning in 1994, was aimed at providing Canadians with affordable access and the necessary skills to use the Internet in primarily rural areas (Cullen, 2001). These skills included basic computer and Internet skills, from elementary Web design to advice about online education (Cullen, 2001). In Canada and the United States, libraries in particular have played a central role in terms of providing computer skills to users (Bertot, Jaeger, McClure, Wright, & Jensen, 2009).

The Economic Opportunity Divide and the Democratic Divide

The **economic opportunity divide** reflects beliefs and attitudes that individuals have about the advantages provided by access, such as finding a job, obtaining health information, and being able to take an online course. The **democratic divide** is the use of the Internet for political engagement, such as obtaining information about political candidates or parties, being able to make donations to political entities, and communicating with government, for example, over email.

Linked to differences in income and education are variations in people's ability to navigate the Internet and find relevant information. Hargittai concludes that

> [m]erely offering people a network-connected machine will not ensure that they can use the medium to meet their needs because they may not be able to maximally take advantage of all that the Web has to offer. Policy decisions that aim to reduce inequalities in access to and use of information technologies must take into consideration the necessary investment in training and support as well. (2002)

As the Web continues to develop and applications are more varied, the access, skills, economic opportunity, and democratic divide is becoming a more pressing policy issue.

The Global Digital Divide

The term **global digital divide** was coined to describe the differences in access to and use of the Internet among nations and regions of the world. Most analysis of the digital divide focuses on disparities within Western nations. An equally pressing concern is, however, the inequality in access to technology between nations. These disparities in access are thought to have major social and economic consequences for nations and their development. The problem is often perceived as one of centre and periphery, where developing countries control the flow of information (Fuchs, 2008). In 2001, the North American dominance was reflected in usage statistics with as much as 60 per cent of the online population being North American (ACNielsen, 2001, as cited in Chen, Boase, & Wellman, 2002). Statistics collected by the United Nations demonstrate that North America, Europe, and Australia have the highest levels of computer penetration with 49 per cent[3] to 89 per cent, which are in stark contrast to the 0 per cent to 5 per cent levels found in some of the poorest regions of the world, including Africa, Central America, and Asia. The low levels of computer usage and Internet penetration in these nations undermine their efforts to improve their citizens' quality of life and reduce their participation in the global economy.

Developing countries continue to struggle in their efforts to become digital, having to overcome numerous barriers that exist at both the micro and macro levels. The most salient ones include the following:

1. *Lack of infrastructure*: The developing world is lacking in terms of hardware, software, and Internet connectivity. This particularly affects rural areas.

2. *Economic barriers*: The cost of computers is one major deterrent for many in the developing world. This cost is coupled with the monthly payment to ISPs that is necessary for Web access.

3. *Illiteracy*: Many parts of the developing world continue to have high levels of illiteracy.

4. *Poor computing skills*: Even if individuals have literacy skills, they may have no previous experience with keyboards or computers.

5. *Lack of support*: It may be difficult to find others who have had past experiences with computers and the Internet to help them troubleshoot computing problems with hardware, software, and Internet navigation.

6. *Cultural barriers*: Often, norms and customs are not in agreement or do not facilitate the use of information and communication technologies (ICTs).

Nonetheless, the composition of Internet users has changed considerably since the early inception of the Web with a large proportion of Asians,

Europeans, and South Americans being avid users. To address the gap between developed and developing countries, the World Summit on the Information Society (WSIS) was organized in 2003 and 2005 by the United Nations with the aim of establishing programs that could target those nations lagging behind. The **International Telecommunications Union (ITU)** published a comprehensive report in 2007 examining the digital divide in 181 economies. To be able to assess advancements in penetration rates, the report categorized economies according to their level of development: (1) **Organisation for Economic Co-operation and Development (OECD) nations**, (2) **developing nations**, and (3) **least developed country (LDC) or nations**.[4]

Figure 7.1 provides charts of penetration rates from 1995 to 2005 for four different kinds of ICTs: fixed lines, mobile cellular phones, Internet, and broadband. In addition, the chart provides information about the gap in technology ownership that exists between countries of different levels of development. The gap is measured as a "ratio of average penetration rates between different groups of countries" (International Telecommunication Union, 2009, p. 22) to further demonstrate the technological inequality that exists.

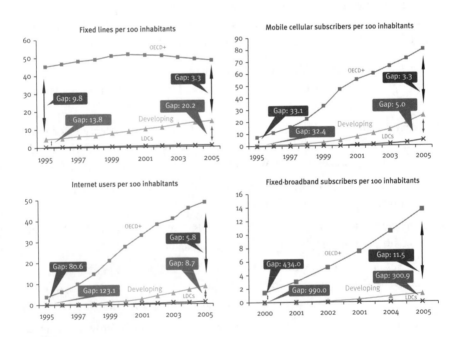

FIGURE 7.1 The Global Digital Divide, 1995–2005
Source: ITU. 2007. World Information Society Report 2007: Beyond WSIS.

While it is certainly encouraging to see the access gap shrinking, we cannot ignore that the divide does persist. We examine next the gaps for each of the four technologies investigated in the *World Information Society Report*.

Fixed Lines

There is evidence that the digital divide for **fixed lines** has shrunk with the gap between OECD countries and developing economies decreasing from 9.8[5] in 1995 to 3.3 in 2005 (Figure 7.1, top left chart). When comparing developing economies with LDCs, by contrast, the gap has considerably widened: in 1995 the gap was 13.8 and in 2005 the gap was 20.2. The data show that middle-income developing countries, in particular China and India, are catching up with OECD countries, while LDCs have stagnated.

Cellphones

The penetration rate for cellphones has increased in OECD countries from about 10 in 1995 to almost 80 in 2005. Data from Scandinavian countries show that penetration rates have reached **saturation**; Norway's penetration rate for 2009 was 111 (International Telecommunication Union, 2009). In these countries, each citizen owns more than one device. A close examination of the top right chart in Figure 7.1 suggests that the gap between OECD and developing countries is rapidly closing from 33.1 in 1995 to 3.1 in 2005. Moreover, even the gap between developing countries and LDCs is narrowing from 32.4 to 5.0. Cellphones are clearly the one technology that can bridge many of the barriers—voice over text, no need for infrastructure, ease of use, and little economic investment—that poor regions of the world are trying to overcome when it comes to connectivity (see Figure 7.1). As a result of these unique features, cellphone technology holds great potential for helping the poorest regions of the world to get connected and participate in the digital economy.[6]

Internet

Internet penetration rates show a different picture. Similarly to cellphones, the penetration rate has increased dramatically in OECD countries: from 5 in 1995 to almost 50 in 2005. By contrast, both developing countries and LDCs are still lagging behind, even though some progress is being made as the reduced gaps show. The gap between OECD and developing countries was 80.6 in 1995 and was 5.8 in 2005, while the gap between developing countries and LDCs was 123.1 in 1995 and was dramatically reduced to 8.7 in 2005. Despite rapid increases in Internet penetration rates in both developing countries and LDCs, the overall penetration rate is still below 10. Therefore, we can conclude on the basis of a close examination of the slope in the bottom left chart in Figure 7.1 that penetration rates are clearly not likely to change much in the near future.

Broadband

The technology divide with broadband is even more apparent than with Internet penetration. The gap between OECD and developing countries was 434.0 in 2000 and 11.5 in 2005. Developing countries show a drastic increase and a slow closing of the broadband divide. In stark contrast, LDCs had a gap with developing countries of 990.0 in 2000 and of 300.9 in 2005. Again, broadband, which is much needed to take advantage of the content, services, and information available on the Web, will only slowly diffuse in developing nations and LDCs. Here is perhaps where the fallouts of the digital divide will become most apparent because these nations will remain in the periphery by not being able to fully participate in the global, digital economy.

Sadly, the digital divide is still a reality and will not dissipate any time soon. As previously mentioned, the barriers to access are numerous and often directly linked to key factors of inequality: poverty, illiteracy, infrastructure, and political/social realities. While ICTs hold much promise, Gurstein points out that they can only have a positive impact if coupled with appropriate socio-economic policy. It is also important to keep in mind that "rather than closing the divide for the sake of it, the more sensible goal is to determine how best to use technology to promote bottom-up development" (*The real digital divide,* 2005). **Information and communication technologies for development (ICT4D)** has aimed to use ICTs directly to reduce poverty and improve health care, education, and work conditions. Box 7.1 discusses the One Laptop per Child initiative as an ICT4D example that has had an impact on some of the poorest regions of the world.

BOX 7.1 ⁕ CLOSING THE GLOBAL DIGITAL DIVIDE: ONE LAPTOP PER CHILD

One Laptop per Child (OLPC) is a US non-profit organization whose mandate is to "create educational opportunities for the world's poorest children by providing each child with a rugged, low-cost, low-power, connected laptop with content and software" (One Laptop per Child, 2010b). The idea behind OLPC is to empower the world's least-advantaged children by involving them in their own education and increasing their connectivity to others.

Nicholas Negroponte, the former head of MIT's Media Lab, established OLPC in 2002 following a trip to Cambodia. There, Negroponte witnessed first-hand the effect of providing children and their families with connected technologies. Understanding that two billion children in the developing world have little or no access to formal education, Negroponte aimed to use new technologies, particularly cost- and energy-efficient computers, to assist in the education of disadvantaged children by providing increased access to resources and tools available through digital sources (One Laptop per Child, 2010a).

Beginning in 2005 with a prototype featuring a yellow hand crank used to power the computer, known as the green machine, OLPC soon began to gain support of organizations such as the **United Nations Development Program (UNDP)**. By November 2006, OLPC had developed its first **XO laptop** using a basic version of Linux's Fedora operating system and an interface known as Sugar.

In spite of the OLPC's good intentions, the program has been heavily criticized. Kraemer, Dedrick, and Sharma have suggested that "expecting a laptop to cause such revolutionary change showed a degree of naiveté, even for an organization with the best intentions and smartest people" (2009, p. 71). While OLPC has presented the XO as a method for helping to close the digital divide, African critics have argued that the scheme tries to impress Westernized ideas of education and progress onto regions where necessities such as clean water, formal education, and health-care programs are of far greater importance (Smith, 2005).

The low cost of the XO laptop should make it affordable for a large segment of the population. However, some have argued that purchasing the device is not so straightforward because OLPC has not taken into account the additional costs associated with the technology in terms of software upgrades and licenses, training, technical support, maintenance, and replacement of broken or malfunctioning computers (Kraemer et al., 2009). Critics have also argued that rather than providing a low-cost alternative to traditional laptops, OLPC instead spurred PC manufacturers to mass-produce small, cheap, and dependable netbooks. In response, companies such as Microsoft started producing affordable versions of their software, and offering them to developing countries and OLPC target markets, such as Nigeria and Libya (Kraemer et al., 2009).

Designed with good intentions, OLPC's program serves as a paradigm of the types of technological and social issues affecting developing countries. In spite of its seemingly minimal startup costs, OLPC's concept overlooked the type of unique cultural, political, and socio-economic conditions facing poorer countries. By viewing ideas of the digital divide through a Westernized lens, OLPC failed to understand the concerns of local environments, particularly in regions where basic nutritional, health, and educational needs trump the purchase and diffusion of digital technologies. Additionally, once OLPC's ambitious program became a competitive threat to traditional industry powerhouses, the originality of their mission became subsumed and overtaken by corporations with greater resources, reputation, and reach in global markets.

The social and economic consequences of the digital divide continue to be the subject of much debate. Chen, Boase, and Wellman argue that

> these inequalities may increase as the Internet becomes more central to global life: from keeping in contact with migrant kin, to acquiring information, to engaging in farm-to-market commerce. Hence, rather than socially including marginal people and countries, the embedding of the Internet in everyday life can enhance and deepen power relations underlying existing inequalities. (2002, pp. 80–81)

China's Move toward Digitization: A Unique Example

The People's Republic of China, or for short China, is a unique case in terms of the diffusion of the Internet for a number of reasons. First, China has seen incredible growth in its information and communications sector—both for personal use as well as business. Second, China has a population of more than one billion, or about 20 per cent of the world's entire population, making it one of the largest countries in the world. Hence, even if only a small proportion of its population is online, it actually represents a large amount of both Internet traffic and, more importantly, Internet content. Third, China is the world's fastest growing economy with a nominal gross domestic product (GDP) of US $4.99 trillion in 2009 according to China's National Bureau of Statistics, making it the third largest in the world after the United States and Japan (Batson, 2010). It is, thus, an important player in the telecommunications sector—both as a consumer and a producer of products and services. Fourth, China's governmental policy has been to promote digitization, while at the same time limiting access to content. Indeed, China has been criticized for its **Internet censorship**, which is enforced through laws, regulations, and repression of citizens. These four examples are just some of the many reasons why China is a unique case in terms of Internet diffusion.

The China Internet Network Information Center (CNNIC) has played an important role in documenting China's digital divide over the past decade and in pointing toward important social and economic trends. A 2006 survey based on telephone sampling found that 123 million users are online, of which 41 per cent are female and 59 per cent are male, and by 2008 that number increased to 298 million. In comparison to other developing countries, the penetration rate in China is fairly high although much lower than in OECD countries. The majority of users are young (see Table 7.2): almost 40 per cent of users are between 18 and 24 years of age. Only 0.8 per cent of users are in the 60-plus age group. Of these users, 26.8 per cent connect via local area networks (LANs), 78 per cent via dial-up, and 77 per cent via broadband. However, insufficient infrastructure in many remote locations and high charges for Internet access are exacerbating the digital divide in China (Flew, 2008).

TABLE 7.2 **Internet Users in China by Age**

Under 18	18–24	25–30	31–35	36–40	41–50	51–60	Over 60
15%	39%	18%	10%	7%	7%	2%	0.8%

Source: China Internet Network Information Center, *18th Statistical Survey Report on the Internet Development in China* (Bejing: CINIC, 2006), p. 9.

As Chinese people embrace the Internet and information flows more freely through China's networks, the government remains ambivalent about join-ing the information society. (See the discussion on the Internet dictator's dilemma in Chapter 8.) On the one hand, China wants to be a key player in the high-tech sector and benefit from the economic advantages provided by e-commerce, networking, and online services. On the other hand, China understands that the Internet's array of communication capabilities, decen-tralized nature, and ease of access to information can potentially undermine the Chinese Communist Party's rule.

The Effects of Technology Transfer on Developing Nations

Through information and communication technologies (ICTs), globalization has reached a new level. The metropolises of the world are interconnected in complex ways through the exchange of goods, services, and information. A central component of globalization is also the transfer of innovations. When a technology is introduced into a new setting, this is referred to as **technology transfer** (Lorimer et al., 2010).

Based on a review of the current literature on technology transfer, Lorimer, Gasher, and Skinner (2010) identify three key insights. First, there is no direct casual link between the introduction of a new technology and resulting social behaviour. As discussed in Chapter 6 in the water boiling example, introducing a technology is a complex social, cultural, and eco-nomic process. While technologies are often introduced by their champions as the solution to an existing problem, they often only provide short-term solutions, not enduring benefits.

Second, many studies of technology transfer tend to focus on the techno-logical advantage of specific equipment and fail to study the socio-cultural system in more depth. They focus on economic returns and less on the con-text in which the technologies will play out. It is, however, important to consider the socio-cultural context as it can provide important information about how the technology may diffuse, change the existing power relations between social groups, and affect social relationships.

Finally, technology transfer results in the receivers of the technology becoming dependent on the suppliers in a number of different ways. Receivers depend on parts, skills, and knowledge to be able to keep technology functioning properly and to be able to repair breakdowns. They start also to depend on newer versions of the technology as older models become obsolete or can no longer be repaired. These kinds of technological dependencies provide economic advantages to the suppliers of technology because machinery, parts, and instruction are purchased. Additionally, unbalanced relationships are formed between those providing and those receiving the technology, as one has power over the other.

These three insights into the consequences of technology transfer help us understand the dynamic relationship between technology integration and society. For those societies that do not develop the technologies themselves, they inevitably fall behind and develop long-term and complex dependencies on the developed nations that supply the technology. Next, we discuss how some of these problems play out in the debate around the digital divide.

Critical Perspectives of the Digital Divide

There is no doubt that the digital divide is a central problem of our times, as it is inextricably intertwined with existing inequalities (Stevenson, 2009). Yet the digital divide concept has also become a tool used by developed nations to impose standards of industrialization on developing countries. Gurstein (2007) provides a thoughtful critique of the digital divide concept and points out four difficulties with research in the area:

1. A large majority of the studies of the digital divide document only the existence of a divide and fail to outline ways of addressing the problem.

2. Most analyses of the digital divide suggest that access alone is a solution to the problem without addressing the larger socio-economic issues that developing countries are struggling with, such as health care, education, and wealth disparities.

3. For Gurstein, the debate is ignoring "the underlying reasons for the impacts of the DD [digital divide] such as on-going trends towards increasing social and economic polarization—with the well-off getting better off and those behind falling even further behind as they find themselves unable to take advantage of ICT opportunities" (2007, p. 45). Providing access will certainly have a social impact on communities, but whether or not access will provide the economic, social, cultural, and educational advantage that lies at the heart of inequality remains unclear. Can access close gaps in inequality?

4. Access needs to be accompanied by training, means of production and distribution, and an economic model of return. Without a more comprehensive plan of development, access will provide little positive return in countries where illiteracy prevails, basic needs are not met, and social unrest continues to exist.

Gurstein's (2007) central point is that the debate surrounding the digital divide should not obscure prevailing inequalities in society, which cannot be easily resolved through technology alone. Similarly, Stevenson (2009) argues that the digital divide is often used as a discursive resource to benefit governments, information capital, and public service professionals. Epstein, Nisbet, and Gillespie (2011) examined how the various definitions of the digital divide had an impact on policy. They found that framing the digital divide as a problem of access emphasized the provision of adequate infrastructure, largely neglecting more fundamental problems of information literacy.

By contrast, framing the problem in terms of skills moves responsibility for the digital divide from governments to citizens, who are perceived as needing to take action in order to obtain the necessary proficiency to access Web resources. This shows that how the digital divide is defined and framed has important repercussions for policy initiatives as the skills perspective may "diminish the public's call for public policies or collective efforts to address the problem," leading people to assume that it is the responsibility of individuals and educational institutions (Epstein et al., 2011, p. 101).

Conclusions

This chapter has provided an analysis of the various attributes and characteristics that define and shape the digital divide. In an era in which the Internet is ubiquitous in everyday life, many would perceive the digital divide to be a relic of the recent past.

Yet despite the rapid consumption and adoption of ICTs, digital inequality remains a pressing issue for both developed and developing nations. The digital divide can no longer simply be correlated with one's ability to access the Internet or own a computer. Rather, the digital divide can be attributed to a lack of individual technical skills to competently use ICTs, or to the inability to seek out information necessary for being an active participant in contemporary politics or culture. Notably, many contemporary scholars and social theorists now view the digital divide in its many overt and distinct forms. These include, but are not limited to, geographical, economic, educational, social, cultural, and technical spheres.

The closing of the digital divide is not simply solved or predicated on supplying poor and isolated individuals with affordable or donated computer hardware and software. It is a project that primarily requires an understanding of the socio-economic factors defining the digital divide, as well as the unique local conditions affecting millions of people who are unfamiliar with the omnipresent and continued experiences of using information technologies as an integral component of everyday life. This issue has both short- and long-term implications, affecting how information is used, manipulated, consumed, and distributed.

Questions for Critical Thought

1. What are the key limitations of early examinations of the digital divide?

2. What are the central barriers to overcoming the digital divide in the developing world?

3. What are the unique challenges that China confronts as it moves into the information society?

4. Provide an in-depth critique of the digital divide concept. In your considerations, address the concept's strengths and weaknesses.

Suggestions for Further Reading

Epstein, D., Nisbet, E.C., & Gillespie, T. (2011). Who's responsible for the digital divide? Public perceptions and policy implications. *The Information Society, 27*(2), 92–104. An investigation into the effect of different digital divide discourses on public perception of responsibility.

Kraemer, K.L, Dedrick, J. & Sharma, P. (2009). One laptop per child: Vision vs. reality. *Communications of the ACM, 52*(6), 66–73. This article critically analyzes the One Laptop per Child program and questions the validity of its solutions for addressing the digital divide among the world's poorest children.

Looker, E.D. & Thiessen, V. (2003). Beyond the digital divide in Canadian schools: From access to competency in the use of information technology. *Social Science Computer Review, 21*, 475–90. The article analyzes access to and use of information technologies in relation to Canadian youth in terms of gender, socio-economic status, and geographic location (rural versus urban).

Norris, P. (2001). *Digital divide: Civic engagement, information poverty and the Internet worldwide.* New York: Cambridge University Press. Examining Internet access in 179 countries, Norris' book looks at how the digital divide affects people's ability to engage in civic life and what its implications are for democracy.

Warschauer, M. (2004). *Technology and social inclusion: Rethinking the digital divide.* Cambridge, MA: MIT Press. Presenting examples from developing and developed nations, Warschauer's book looks beyond the simple view of the digital divide as being purely equated to Internet access, or to the lack thereof, to how the digital divide affects social and economic stratification and inclusion.

Witte, J.C., & Mannon, S.E. (2010). *The Internet and social inequalities.* New York: Routledge. Using sociological principles, this book investigates how use of the Internet reflects and assists in shaping contemporary social inequalities.

Online Resources

Internet Corporation for Assigned Names and Numbers (ICANN)
www.icann.org/
  ICANN is an international non-profit partnership to keep the Internet secure, stable, and interoperable. The website contains information on the history, mission, and current challenges of managing unique Internet identifiers or addresses.
United Nations Information and Communication Technology.
http://stdev.unctad.org/themes/ict/docs.html/
  The website contains reports from around the world about the current state of the digital divide as well as information on recent conferences addressing concerns around the issue.
World Internet Project (WIP)
www.worldinternetproject.net/
  WIP is a major international, collaborative project founded in 1999 by the University of Southern California Annenberg School's Center for the Digital Future, and is focused on examining the social, political, and economic influence of the Internet and other new technologies. The website contains publications on household and nation adoption and use of the Internet.

8 Community in the Network Society

Learning Objectives

- ⊛ To investigate "the community question" and its link to industrialization.
- ⊛ To provide an overview of the social capital concept and its relevance to community.
- ⊛ To discuss and critically examine how the Internet has increased or decreased the social capital available in communities.
- ⊛ To re-examine the concept of the public sphere in light of widespread use of information and communication technologies, in particular social media and cellphones.

Introduction

In this chapter, we examine the impact of technology on social structure. The chapter starts with a brief overview of definitions of *community* and introduces two key theoretical concepts that help readers to better understand the basic structuring of society—*Gesellschaft* and *Gemeinschaft*. What follows is a critical examination of the debate over how community changed as a result of industrialization, urbanization, and globalization by comparing three different, prevalent perspectives: community lost, community saved, and community liberated. The chapter also introduces the term *social capital* to better explain how resources are mobilized within communities. We review two types of social capital—private and public—and argue that each contributes in a unique way to the well-being of communities. Next, we discuss recent theorizing that suggests that social capital is in decline, and we consider the consequences resulting from this decline. One of the key factors identified as affecting this decline is television.

The chapter also considers the impact of the Internet on community and presents various competing perspectives. On the one hand, early analysts of the Internet characterized it as a utopian place, where new communities of solidarity could be formed without constraints of space and time. On the other hand, skeptics saw the Internet as another technology that would draw people away from family and friends and further alienate them from society. We contrast and critically examine the utopian and dystopian views, arguing that we need to develop new perspectives to better understand how the Internet has affected social structure. The chapter concludes with a discussion of how the Internet has affected our understanding of the public sphere. As part of this discussion, we take an in-depth look at the events that unfolded in Egypt in February 2011 and analyze the role that social media played in initiating, supporting, and helping to organize the protests.

What Is Community?

We can study communities from a number of different perspectives. Traditionally, researchers have studied communities in terms of location; that is, a community is a group of people who live in a bounded geographic area. However, we can also study community in terms of smaller social units that come together because of shared interests, work, religion, etc. By contrast, some very broad definitions include all of society as community.

Ferdinand Tönnies (2004) was the first to study the fundamentals of community and proposed the distinction between **Gemeinschaft** and **Gesellschaft**. *Gemeinschaft* is generally translated as "community" and refers to a cohesive social entity that is united by a pre-existing bond. *Gemeinschaft*-based affiliations are ends in and of themselves and do not directly serve a utilitarian purpose even though benefits can be obtained from the association. Family ties are a perfect example of a form of *Gemeinschaft* that connects people in tightly knit groups. By contrast, *Gesellschaft* is translated as "society" or "association," describing the coexistence of individuals who are self-serving units and come together because of an over-arching goal. In *Gesellschaft*-based associations, people are only loosely connected through bonds that are often goal-oriented. The **nation-state** is an example of *Gesellschaft* because its members are grouped together as a result of sharing the same geographic boundaries and national identity.

Gemeinschaft and *Gesellschaft* represent ideal types or archetypes of social relationships, that is, most social groups cannot be categorized into either pure *Gemeinschaft* or *Gesellschaft* but tend to have elements resembling one or the other. Tönnies (2004) shows through his analysis of social life that both forms of social organization—*Gemeinschaft* and *Gesellschaft*—can coexist at a single point in history as each describes a different aspect of social organization. Individuals live in narrowly defined groups based on kinship, location, and affiliation (e.g., religion), but at the same time can be part of larger social structures that attain goals (e.g., nation-states, jurisdictions).

Gemeinschaft is often used to represent the ideal type of society, one where people are closely connected, share an identity, and engage in reciprocity. This contrasts with *Gesellschaft*, which is perceived as an inferior form of social organization because members are alienated, there is little willingness for co-operation and collaboration, and people live segregated from each other. Next, we focus our discussion on how industrialization, urbanization, and bureaucratization have affected community.

Community's Link to Industrialization

Barry Wellman has used the term "the **community question**" to summarize the debate around how community has changed over time and the role played by technology. He argues that this is a pressing issue because it links micro- and macro-level analysis in that it addresses "the problem of the structural integration of a social system and the interpersonal means by which its members have access to scarce resources" (Wellman, 1979, p. 1201). The central question has been whether communities are declining or thriving. Wellman (1979) distinguishes between three theoretical views: community lost, community saved, and community liberated.

Analysts who belong to the **community-lost view** have painted a bleak picture of the state of community in Western societies. They see industrialization as the cause of a decline in community, resulting from long work hours that leave little time for other activities. For instance, women have moved in large numbers into the workforce with increasingly long work hours, and this has changed how socialization takes place in the home and neighbourhood (Costa & Kahn, 2001). Urbanization, in conjunction with urban sprawl, creates isolation and a general lack of public spaces, further reducing opportunities for socialization. New modes of transportation and communication have emerged that support distant interactions, removing people from their immediate vicinities and, ultimately, creating loose-knit communities.

At the centre of the community-lost argument is the comparison between contemporary urban living and pastoral community. Supporters of the community-lost perspective see pastoral community as composed primarily of local social interactions in closely bounded groups. In these communities, people are primarily involved with fellow members of the few groups to which they belong: at home, in the neighbourhood, or at work. These networks are fairly homogenous, with members sharing a common geography, identity, and belief system, as well as common social ties. In part, people idealize pastoral community by portraying it as a simple but happy way of life.

The **community-saved view** was developed in opposition to the community-lost view. This perspective focuses on how friendship and family networks continue to dominate as forms of social organization even in heavily industrialized societies. Part of the argument is that a move is taking place where loosely bounded networks often increase their level of connectivity, leading to close-knit clusters similar to those found in the pre-industrial era. The evidence has consistently shown that close-knit groups continue to exist, in particular within poorer neighbourhoods, where people need to rely on each other for emotional, economic, and social support. Hence, despite the social changes taking place as a result of industrialization, urbanization, and modernization, there is some evidence supporting the idea of community saved.

The community-lost and community-saved views, however, have several limitations and therefore require some rethinking. Wellman has criticized the community-saved view because of its narrow focus on documenting the continued existence of pre-industrialized forms of community without giving much consideration to the changes that have occurred in the structure of society since the Industrial Revolution (Wellman, 1979). In addition, Wellman criticizes both community-lost and community-saved views because of their disregard of the social networks that develop and form outside geographic boundaries:

> Thus the basic Community Question, dealing with the structure and use of primary ties, has been confounded in both the Lost and Saved arguments with questions about the persistence of solidary sentiments and territorial cohesiveness. But, whereas the Lost argument laments their demise, the Saved argument praises their persistence. (p. 1206)

Wellman's comment shows that the community-lost view tends to focus on the loss of geographically bound ties without giving much consideration to social ties formed outside the core groups of the neighbourhood and family. By contrast, the community-saved view tends to disregard changes that have occurred in the structuring of society as a result of modernization by emphasizing the continued existence of geographically bounded ties.

A third perspective is depicted in the **community-liberated view**, where the central argument is that community life is not lost but has undergone radical transformations. In this view, communities continue to exist in society but with new dimensions. Instead of socialization taking place within narrowly defined geographic boundaries, people socialize outside of local neighbourhoods and family ties (Guest & Wierzbicki, 1999; Rainie & Wellman, 2012; Wellman & Frank, 2001). Indeed, while immediate neighbours may not know each other or may not socialize as frequently as those in the pastoral communities of the eighteenth century, socialization does continue with friends and family who live at a distance (Fischer, 1992; Mok, Wellman, & Carrasco, 2010).

The car and the telephone facilitate the formation of these new social structures, where emphasis is placed on establishing and maintaining unbounded networks (see Chapter 9). While in-person visits continue to be the primary form of socializing, the telephone occupies a unique role in that it promotes distant communication (Wellman & Wortley, 1990).

The community-liberated view provides unique insights into the structuring of society and the link between micro- and macro-level developments.

Clearly, industrialization, urbanization, and bureaucracy have left their marks on society. What is relevant, then, is to understand what the nature of community is in the context of modern life.

Social Capital and Its Relevance to Community

To further explore the nature of community in the context of modern life, this section focuses on social capital. **Social capital** is a useful term because it describes an individual's or a group's actual and potential access to valuable resources through social networks. Social capital, thus, is the sum of valuable resources that can be obtained through the relationships actors have with friends and relatives and the **social networks** that these relationships form (Wellman & Berkowitz, 1988). The focus in the social capital perspective, then, is no longer only on the formation of community but also on how members of a community manage their resources. These resources are of most importance when building a healthy community because, according to Putnam, "[a] growing body of literature suggests that where trust and social nets flourish, individuals, firms, neighborhoods, and even nations prosper" (2000, p. 319). In communities, where social capital is high, trust among individuals is high, as is reciprocity. That is, people give without expecting immediate returns, in turn creating an atmosphere of mutuality. Therefore, we must understand the concept of social capital and how it can be enhanced.

We cannot trace the origins of the term *social capital* to a single individual (Borgatti, 1998; Putnam, 2000). In fact, researchers in different countries and research areas simultaneously used the term. However, Jane Jacobs, a renowned Toronto-based writer and social activist, was an early user of the term. She is known for her participation in the 1960s social movement that stopped the construction of a controversial highway project. The highway, known as the Spadina Expressway, was to travel through the city and divide a number of neighbourhoods. Jacobs was interested in understanding how grassroots efforts could be organized at the community level in order to represent a community's concerns to city council. For her, people's relations and the networks they formed were essential as "[t]hese networks are a city's irreplaceable social capital. Whenever the capital is lost, from whatever cause, the income from it disappears, never to return until and unless new capital is slowly and chancily accumulated" (Jacobs, 1961, p. 138). Hence, a community needs to value and properly manage its social capital to continue to prosper.

We can view the benefits derived from social capital as either private or public effects. Private effects are benefits brought to individuals through their ties, while public effects are all those positive characteristics of living in densely knit and reciprocal communities. For example, a person finding a job with the help of a friend reaps the benefits of private effects, whereas feeling safe at night on the streets because neighbours are watching out for each other is an illustration of the advantages accrued through public effects.

Another benefit of social capital is that it facilitates the flow of resources because it is based on norms of **generalized reciprocity**, which is expressed in the notion of doing favours for and providing help to others without expecting them to return the favour immediately (Putnam, 2000). The underlying assumption is that at a future point in time when one is in need, the favour will be returned and not necessarily by the same person.

TV and the Decline of Social Capital in the United States

A major concern has been a decline in social capital in the United States, which can have serious implications for community, solidarity, and ultimately the vitality of a democratic society (Putnam, 2000). To document the decline, Robert Putnam examines the extent to which Americans were involved in a number of social behaviours, including whether or not they (1) attended church services, (2) visited relatives, (3) gave or attended a dinner party, (4) attended a sports event, and (5) visited neighbours. Using US General Social Survey (GSS) data, Putnam demonstrates that people are less engaged in their community now than they were in the 1960s and 1970s. For instance, Figure 8.1 shows a steady decline in informal socializing, that is, visiting friends, attending celebrations, visiting bars, and participating in informal conversations.

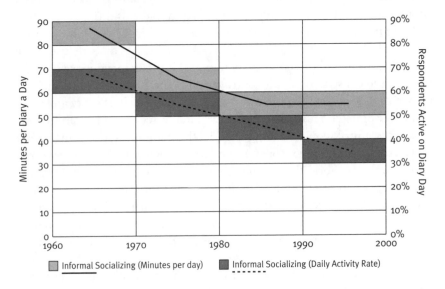

FIGURE 8.1 Informal Socializing, 1965–1995
Source: Adapted from Putnam, R.D. (2000). *Bowling Alone: The Collapse and Revival of American Community.* New York: Simon and Schuster, p. 108.

Each social or community activity can be seen as an opportunity to increase private and public social capital. Through these interactions, a person can learn about job opportunities and information about the community. Therefore, the social capital obtained through formal as well as informal relations is relevant for the prosperity of individual members as well as for the community as a whole.

For Putnam (2000), technology has directly contributed to this decline. He argues that with industrialization came increased participation in more individualistic activities, such as watching television. The move toward individualized activities has had an enormous impact on how people socialize and participate in their communities. Entertainment tends to occur more within the home, reducing opportunities for socialization, for meeting neighbours, and for getting involved in community activities. Putnam (2000) further argues that watching television is a passive form of entertainment that does not involve chatting, debating, or interacting. In addition, television has an absorptive effect that even reduces social exchanges among family members in the home, further contributing to the demise of social capital (Steiner, 1963). Although television has indeed had a considerable effect on socialization, scholars see the impact of the Internet as more pronounced and transformational.

Revisiting Community in the Internet Era

What impact has the Internet had on community? Has the Internet strengthened community and increased social capital? Or has it further isolated individuals and decreased community involvement? Since the 1990s, there has been an ongoing debate about how the Internet has affected socialization, communication, and civic participation. We discuss three competing perspectives previously proposed by Wellman, Quan-Haase, Witte, and Hampton (2001): utopian, dystopian, and supplement.

In the **utopian perspective of the Internet**, analysts see the Internet as stimulating positive change in people's lives. According to this perspective, the Internet has changed our concept of community because it spans geographic boundaries and connects individuals across time and space. As well, digital media are perceived as leading to the formation of new forms of community that allow for people with common interests to meet.

Rheingold defines **online communities** as "social aggregations that emerge from the Net when enough people carry on those public discussions long enough, with sufficient human feeling, to form webs of personal relationships in cyberspace" (2000, p. xx). Rheingold based his definition on his own experience in a virtual community called the **WELL** (Whole Earth 'Lectronic Link), which started in the 1980s and had a large following in the San Francisco Bay Area. The online community formed a close-knit social network that provided members with friendship, social and emotional support, information, and a digital space for discussion.

From a utopian view, the social effects of the Internet apply to many social realms, including e-democracy, e-learning, and e-health. In stark opposition to the utopians is the dystopian perspective, which we discuss next.

The key argument of the **dystopian perspective of the Internet** is that the Internet draws people away from their immediate, local environments, potentially alienating them from social engagement and civic participation. Theorists continue to raise concerns about how technology is making us lonelier. Early theorists focused on the Internet's anonymous nature, the numerous possibilities for deception, and the rather shallow relationships that are formed online. Dystopians argued that in-person communication was richer, more fulfilling, and provided interactivity, whereas text-based exchanges online were alienating. Putnam (2000) argues that online interactions, like television viewing, can be immersive, taking people away from their immediate in-person contacts. Another concern Putnum raised was the negative effect that global connections could have on local community. Furthermore, the Internet could have detrimental effects on the public sphere as people move to the Internet for entertainment and socializing and move away from public spaces (Putnam, 2000).

The utopian and dystopian perspectives both provide rather simplistic views of the impact of the Internet on the structuring of society. Perhaps these perspectives have given too much weight to the Internet in terms of its ability to radically transform the nature of community. While the Internet can provide the means for increased interaction and community involvement, it can also isolate individuals and increase the risk of harm. To better understand how the Internet has affected society, we need to consider a number of mediating factors:

- a user's previous experience with the Internet;
- a user's personal characteristics (e.g., age, gender, and personality);
- the existence of prior forms of community, online or offline; and
- the type of Internet use (e.g., surfing, gaming, emailing, or chatting).

Only by carefully analyzing these factors will we be able to better understand how the Internet impacts society.

Overall, the evidence suggests that the Internet provides an important and central means of communication. However, as a medium it *adds* to other forms of communication rather than replacing them. This third perspective has been termed the **supplement argument of the Internet** (Wellman et al., 2001). Box 8.1 reviews a study that provides evidence in support of the supplement argument.

BOX 8.1 ⁂ SOCIAL CAPITAL IN THE INTERNET ERA

Evidence suggests that the Internet neither decreases nor increases **social capital**. A number of large-scale studies have demonstrated that with an increased reliance on the Internet for socialization and communication, the use of other media does not decrease (Howard, Rainie, & Jones, 2002). This line of thinking has been referred to as the supplement argument (Wellman et al., 2001).

In a study of US and Canadian Internet users, email had become an important means of keeping in touch with friends and family (Wellman et al., 2001). However, as the amount of email sent and received increased, interactions and phone calls did not decrease. This strongly suggests that email supports existing social networks, but does not replace telephone and in-person communication. Indeed, Table 8.1 shows that people continue to use the telephone the most for contact with friends and family regardless of geographic proximity. This is followed by email for friends both near and far.

In contrast, face-to-face meetings continue to be used more frequently than email exchanges for communicating with family members who live nearby. People use email to keep in touch with family who live far away because of distance constraints. Interestingly, individuals who have little social contact via telephone or in person are also unlikely to use email for socializing. Similarly, people who visit and telephone frequently also email frequently, suggesting that those who already have established communication patterns carry these seamlessly over to the Internet.

TABLE 8.1 **Social Contact with Friends and Family, Near and Far**

	Phone (Days/Year)	F2F (Days/Year)	Email (Days/Year)	Letters (Days/Year)
Friends Near	126	92	118	9
Family Near	114	58	49	7
Friends Far	25	10	85	8
Family Far	43	10	72	10

F2F = face-to-face
Source: Huysman, M., and Wulf, V. (Eds.). Social Capital and Information Technology, Figure 8.1, Social Contact with Friends and Family, Near and Far, © 2004 Massachusetts Institute of Technology, by permission of The MIT Press.

The data show that the Internet has joined the telephone and in-person communication as a main means of staying in touch—but one that can be more convenient and affordable. The Internet offers a new array of communication possibilities, including real-time communication via chat, Twitter, Facebook, and blogs (Klemens, 2010). Clearly, this affords new possibilities in terms of the ease in connecting geographically dispersed people and organizations bonded by shared interests (Quan-Haase & Wellman, 2004). In sum, the capabilities of the Internet add to and supplement interactions with other media rather than replacing them. Not all pre-established social behaviours have been revolutionized; people continue to visit friends and family in person and to talk on the phone.

The evidence so far suggests two trends emerging in how the Internet has affected community, social networks, and communication: (1) the rich get richer hypothesis and (2) networked individualism. The **rich get richer hypothesis** argues that the Internet does not have the same effect on all users; instead, for those users who are socially involved and who obtain social support, the Internet will further strengthen and expand these networks (Kraut, Kiesler, Boneva, Cummings, Helgeson, & Crawford, 2002; Tufekci, 2010). The Internet will benefit these people because it will provide an additional medium to keep in touch. Kraut et al. argue that "[t]hose who already have social support can use the Internet to reinforce ties with those in their support networks. If so, these groups would gain more social involvement and well-being from using the Internet than those who are introverted or have limited networks" (2002, p. 58). Furthermore, the rich get richer hypothesis predicts that those who already have large networks can use the Internet to both maintain these networks more efficiently and to continue increasing their network size.

Networked individualism, on the other hand, argues that society has moved away from a model where people are embedded in groups (*Gemeinschaft*) to more loosely connected networks (*Gesellschaft*) (Rainie & Wellman, 2012; Wellman, 2001). Individuals no longer feel a strong commitment to groups but instead tend to build and maintain their own personalized networks (Rainie & Wellman, 2012). Barry Wellman writes that "This is a time for individuals and their networks, not for groups. The all-embracing collectivity (Parsons 1951; Braga and Menosky 1999) has become a fragmented, personalized network. Autonomy, opportunity, and uncertainty rule today's community game" (2001, p. 248).

Social media facilitate a networked structure of interaction because they allow each participant to maintain his or her individualized social network. For example, Facebook allows each user to add friends, family, and acquaintances without any restraints of group affiliation. While networks in digital space may overlap, they are still distinct enough to argue that they represent a social structure where people move over time between various groups and networks, instead of being a member of a single network. This is a fundamental shift in the structure and functioning of society.

How Is Technology Transforming the Public Sphere?

To what extent is technology eroding the existence of the **public sphere**? The public sphere is essential for democracy because it provides "a discursive space in which individuals and groups congregate to discuss matters of mutual interest and, where possible, to reach a common judgment. Public spheres are discursive sites where society deliberates about normative standards and even develops new frameworks for expressing and evaluating social reality" (Hauser, 1998, p. 86). While public spaces—such as plazas, parks, and centres—were at the heart of the public sphere in the past, an increasing reliance on technology means that these places have become less relevant. Urbanization, industrialization, and the spread of suburbs have all contributed to the erosion of the public sphere. Recent debate has focused on how the introduction of new digital technologies will transform the public sphere. Next we discuss various approaches developed to better understand this complex interrelation.

Theories of the Demise of the Public Sphere

Sunstein (2001) has warned about the problems that arise from using technology to access information. While the advantage of technology is that using it allows people to filter information and customize their selection, this advantage at the same time limits people's exposure. Because the Internet allows users to visit websites that are very specialized and often geared toward specific audiences, the Internet eliminates an element of randomness, reduces exposure to a variety of views and perspectives, and potentially creates a biased world view. For Sunstein (2001), digital technologies can lead to reducing information in the public sphere, thereby creating a new form of society, which he terms Republic.com, because these tools make it easier to filter, personalize, and customize content.

A similar trend has been documented with the use of iPods (also see Chapter 2). Bull (2008) has provided the most comprehensive social analysis of the iPod and how it links individuals to the public sphere. According to Bull, users filter information available in their environment through their self-submergence in the iPod. The iPod "is symbolic of a culture in which we increasingly use communication technologies to control and manage our experiences in the urban environment" (Bull, 2008, p. 4). The iPod enables people to carve an individual sense of **personal auditory space** during commuting and other activities, thereby separating themselves from their surroundings. We observe a further move toward technologizing travel: the car, train, and airplane have already transformed how we move through space, and with the iPod this trend is taken a step further in that commuters can now "create private domains to withdraw into while being on display in the public realm" (Boradkar, 2006, p. 27).

The analyses by Sunstein and Bull together suggest that digital technologies are a contributing factor in the demise of the public sphere. Instead of connecting citizens and facilitating conversation, debate, the flow of information, and the organization of political and social movements, digital technologies are contributing to narrow perspectives and are further isolating citizens. Next, we explore a different angle of the debate about how the Internet affects the public sphere.

Third Places

Another important element of the public sphere is the notion of the **third place**. For Oldenburg (1999), the home is the first place, where family and friends come together. Work is the second place, where people spent a lot of their time and have a distinct set of co-workers and friends. Third places are coffee houses, taverns, restaurants, bars, libraries, and other locations that people visit routinely. These places ground people in the neighbourhood and community and allow them to interact and develop a sense of place. Third places are important locations for public opinion to form and for civil society to thrive.

How do digital technologies impact the role of third places in our technological society? Putnam (2000) argues that television has pulled people away from third places, with entertainment now taking place within the walls of the home. From his viewpoint, the Internet will have similar effects, immersing people in a world of information that has little overlap with their local community.

It is true that the data show fewer people engaging in local organizations, attending church, and participating in politics, as discussed earlier; however, digital technologies are in fact playing a *different* role in creating third places and rebuilding the public sphere. Digital technologies seem to be filling the gap left by traditional third places by providing alternative spaces to hang out, meet people, exchange ideas, debate social and political topics, and post and exchange information. While television seemed to contribute to the erosion of the public sphere (Putnam, 2000), the Internet as a medium provides much greater capacity for citizen engagement. (See the discussion of produsage in Chapter 6.)

The Internet as a Tool of Political Engagement

Indeed, the Internet is not only a source of entertainment but also a tool for political engagement and for civil society to emerge. The interactive nature of digital communication allows users to engage with material in a different manner than what TV affords. Not only can debate arise online, but community can form around political, social, and economic issues of concern to citizens. We will discuss this in Box 8.2, where we consider the recent uprising in the Middle East and the role social media has played in this particular social movement.

Box 8.2 ❋ SOCIAL MEDIA IN THE MIDDLE EAST

Social media has taken off in the Middle East with a large increase in users (Social Bakers, 2011). For example, protests in the Arab region started not in the streets but with short tweets on Twitter and posts on Facebook. The recent uprising in Tunisia and Egypt has been labelled a **Twitter Revolution** because the Internet provided a platform to organize, mobilize, and voice opinions. In an eerily accurate prediction of the soon-to-take-place Twitter Revolution, Philip Howard, in his book *The Digital Origins of Dictatorship and Democracy* (2011), titled the prologue "Revolution in the Middle East Will Be Digitized." Howard's prescience shows how ICTs have become an integral part of political and social debate in the Middle East, adding to traditional forms of exchange taking place in coffee and tea shops, in taverns, on the streets, and in markets.

As the tension on the streets of Cairo increased, the government felt pressure to act quickly to evade a digital revolution. Its strategy was to shut down the Internet in an attempt to freeze the masses and stop them from organizing and going to the streets. This shutdown has been referred to as the use of the **Internet kill switch** and has opened a debate about the possibility of governments using their power and legislation to shut down the Web. The Egyptian government correctly recognized the power of social media to influence people to protest. The movement was not a small group of people in isolation rallying for change but, rather, the digital sphere mobilizing for social and political change. While the government identified the role of social media in initiating the protests, it made a mistake in thinking that flipping the Internet kill switch would stop the protests. The government not only cut citizens off from current national and international information about how the events were unfolding, but it also halted the protesters' ability to communicate and organize. In addition, the government cut off citizens from their cellphones, which meant no calls, no texting, and no Web access. Egypt was basically a zone of zero information and connectivity.

Anti-Mubarak protester holding a sign praising Facebook for helping to organize the 2011 protest in Tahrir Square, Cairo, Egypt.
Source: © Idealink Photography/Alamy.

Surprisingly, this isolation did not stop the movement but, rather, gave it further momentum. Now that people were cut off from all telecommunications, they went to the streets to express their anger and discontent about what they felt was an abuse of power on the part of their government. As a result, this change in government strategy infused a social movement with renewed strength and focus.

A key issue has been the role of the Internet both before and during the uprising (Howard, 2011). The media have been quick in concluding that this revolution was caused by social media as the term "Twitter Revolution" suggests. However, analysts have cautioned people about making conjectures about how social media has affected the process. Indeed, Zuckerman writes about how other factors have been central in leading up to the uprising: "But any attempt to credit a massive political shift to a single factor—technological, economic, or otherwise—is simply untrue. Tunisians took to the streets due to decades of frustration, not in reaction to a WikiLeaks cable, a denial-of-service attack, or a Facebook update" (2011).

Even though for Zuckerman social media is not the single factor in creating social change, he acknowledges that social media played an intrinsic role in how the protests unfolded: "But as we learn more about the events of the past few weeks, we'll discover that online media did play a role in helping Tunisians learn about the actions their fellow citizens were taking and in making the decision to mobilize" (2011). Some analysts argue that social media were particularly critical in enabling protest leaders to mobilize; they are referred to as **digital revolutionaries**. Wael Ghonim, who is a Google executive and played a role in the early stages of the uprising, saw great value in social media: "the revolution started on Facebook," and "if you want to liberate a society just give them the Internet" (MacKinnon, 2011). The Arab Spring movement used social media extensively to voice opinions, exchange information, and organize during the early stages of the protests. Figure 8.2 depicts the number of followers of Ghonim's Twitter account as the events unfolded. We observe a stark increase between 6 and 11 February from about 20,000 followers to as much as 80,000.

Srinivasan argues that we cannot deny the impact the Internet is having on politics. He writes, "With four billion mobile phone users and 30% of the world's population with basic Internet access, it's absurd to dispute the implications of these technologies on social, political, and economic life" (2011). Even individuals who are not connected are nonetheless affected by the technological changes. The Internet in and of itself is not a political tool; as Table 8.2 shows, people can use social media for many different

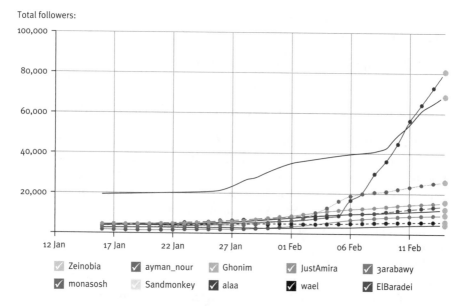

Total followers:

FIGURE 8.2 Ghonim's Followers on Twitter
Source: Tufekci, Z. Can "Leaderless revolutions" stay leaderless: Preferential attachment, iron laws and networks, *Technosociology* blog, 14 February 2011. Retrieved from http://technosociology. org/?p=366.

TABLE 8.2 The Uses of Social Media

Medium	Personal Use	Use for Activists
YouTube	Cute videos of your cats	Videos of trade union protests
Flickr	Cute photos of your cats	Subversive photos unblocked by firewalls
Google	Overlay your cute cat photos on a map of your neighbourhood	Overlay of prisons, land ownership in your country
Twitter	Real-time updates on your cats being cute	Real-time updates on whether activists are free or imprisoned
Blogger	Links to aspects of your cats' online presence	Online newsrooms reporting on activist activities

Source: Based on Zuckerman, E. The cute cat theory talk at ETech," *My Heart's in Accra* blog, 8 March 2008. Retrieved from www.ethanzuckerman.com/blog/2008/03/08/the-cute-cat-theory-talk-at-etech/.

reasons. What is unique about social media, however, is their ability to connect people in real time in an organic and networked structure, which requires little centralization and command. New ICTs "such as the internet and mobile phones, have had clear roles in both starting new democratic processes in some countries and in entrenching them in others" (Howard, 2011, p. 3).

The term **digital public sphere** describes the new forms of association that are developed online and the possibilities that they provide for citizens to organize and mobilize. Some of the central changes that have occurred as a result of the digitization of the public sphere according to Gripsrud, Moe, and Splichal (2010) are as follows:

1. Changes and processes occur on a global scale.

2. Various media converge and blur.

3. There has been a movement toward the commercialization of the media.

Hillary Rodham Clinton, in her February 2011 speech on "Internet Rights and Wrongs: Choices & Challenges in a Networked World," portrayed the Internet as the public sphere of the twenty-first century: "The Internet has become *the* public space of the 21st century. The world's town square, classroom, marketplace, coffeehouse and nightclub. We all shape and are shaped by what happens there. All two billion of us and counting" (2011, emphasis in original).

Clinton has also addressed the challenge the Internet poses to be both an open space for debate as well as a source of economic growth. This challenge, often referred to as the **Internet dictator's dilemma**, creates enormous pressures on oppressive regimes. At the core of the problem is the push toward greater integration into the digital economy through e-commerce, e-health, e-education, and other areas with simultaneous control over Internet use and censorship of content. Clinton has pointed out that "They will face a dictator's dilemma, and will have to choose between letting the walls fall or paying the price to keep them standing" (2011).

As Day and Schuler have noted, the "increasing communication and collaboration between social movements, civil society and community networks does . . . possess the potential for an emerging counter-culture" (2006, p. 20). Many of these initiatives are created and organized from the bottom up, through the use of a range of participatory tools that empower the community (Day & Schuler, 2006). These communities are diverse and vary in purpose: "[t]hey are not like organisational structures—the boundaries of which can be identified, quantified and measured—communities are messy, hard to pin down and problematic" (Day & Schuler, 2006, p. 27). Our discussion of the digital public sphere concludes with a comment by Srinivanas: "Even if new technologies can serve both democratic and repressive purposes, no one disputes their continued growth as the economic, social, political, and cultural substrate of our times" (2011, n.p.).

Conclusions

Community is in constant flux. As social, technological, and economic change take place, community and its forms of expression also shift. While much of the debate around the concept of community has focused on how much it has changed from the early pastoral community of the eighteenth century to our current high-tech society, this comparison does not really address the more important issue of what the current forms of community are. As people change their communicating and socializing patterns, new theories and measurements of community need to develop as well.

Early writings emphasized either the Internet's negative or positive impact on social structure. The utopian perspective saw new communities of solidarity forming online without constraints of space and time. By contrast, the dystopian perspective emphasized potential negative effects that resulted from text-based communications that did not afford the rich contextual forms of exchange made possible in person. Both perspectives tend to provide a limited understanding of the impact of the Internet on the structuring of society. The Internet is blending into the rhythms of everyday life, supporting mundane activities such as banking, surfing for information, planning vacations, etc.

The Internet is also creating a sphere for new forms of socialization. Facebook and Twitter provide new platforms for staying in touch with large numbers of friends. These tools are changing the structure of society—people are no longer embedded in narrowly defined, local clusters, but instead tend to socialize with a heterogeneous, global social network. These networks are characterized by constant change as people shift locations, interests, and jobs. This new form of structure has been referred to as networked individualism because each person has his or her own unique network of family, friends, co-workers, and acquaintances. Overall, the evidence suggests that industrialization, urbanization, bureaucracy, and the digital media did not destroy community but instead have transformed its structure, composition, attitudes, and communication practices.

The public sphere is also undergoing transformations as a result of digital technologies, which provide alternative means of voicing, sharing, and debating opinions. Despite these new opportunities, the use of social media also creates new challenges. For example, in many countries using Facebook as a means to express dissenting and alternative opinions about controversial topics circumvents censorship and oppression. From this, new challenges emerge as Facebook and other social media need to provide secure environments where dissidents can express their views without jeopardizing their safety (MacKinnon, 2011). Social media is still in its infancy and as its use becomes more integrated into society, both opportunities and challenges will arise.

Questions for Critical Thought ....

1. Does the Internet create a social structure that resembles *Gemeinschaft* or *Gesellschaft*?

2. Provide an overview of the evidence that suggests the Internet is neither increasing nor decreasing social capital.

3. Define "Internet kill switch" and discuss whether or not you think governments should have a right to shut down the Internet.

4. Discuss the role of social media (Twitter, Facebook, and YouTube) in the recent uprising in Tunisia and Egypt.

Suggestions for Further Reading

Castells, M. (1996). *The rise of the network society.* Cambridge, MA: Blackwell Publishers. In this comprehensive book, Castells shows how society has fundamentally changed as a result of information and communication technology.

Howard, P.N. (2011). *The digital origins of dictatorship and democracy: Information technology and political Islam.* Oxford: Oxford University Press. An insightful book on the changes taking place in the Middle East and the role of information technology.

Putnam, R.D. (2000). *Bowling alone: The collapse and revival of American community.* New York: Simon and Schuster. The author argues that social capital is decreasing in America and shows how technology is linked to this decline.

Wellman, B., Quan-Haase, A., Witte, J., & Hampton, K. (2001). Does the Internet increase, decrease, or supplement social capital? Social networks, participation, and community commitment. *American Behavioral Scientist, 45*(3), 437–456. This article is a classic and presents data on the impact of the Internet on social capital.

Online Resources

Digital_nation: Life on the virtual frontier
www.pbs.org/wgbh/pages/frontline/digitalnation/etc/synopsis.html
    This outstanding PBS website for the program *Frontline* features PBS documentaries and interviews that examine what it is like to live in a digital world.

My Heart's in ACCRA
www.ethanzuckerman.com/blog
    An insightful blog maintained by Ethan Zuckerman, a senior researcher at the Berkman Center for Internet & Society.

International Network for Social Network Analysis (INSNA)
www.insna.org/
    INSNA is an association that brings together parties interested in methods and theory of social network analysis.

US Secretary of State Hillary Rodham Clinton Speaks on Internet Freedom

www.youtube.com/watch?v=ccGzOJHE1rw or for text http://www.cfr.
org/democracy-and-human-rights/clintons-speech-internet-freedom-
january-2010/p21253
This is a landmark speech on the role of the Internet in global civil
society.

9 Technology–Mediated Social Relationships

Learning Objectives

- ⊛ To provide a historical overview of the development of mediated communication and its impact on society.
- ⊛ To present current ideas on how the meaning of the term *friendship* has changed as a result of social media.
- ⊛ To investigate how the Internet has affected the formation and dissolution of romantic relationships.
- ⊛ To explore the concept of virtual mourning and how it shifts our understanding of death.

Introduction

In this chapter, we briefly outline the early beginnings of mediated communication and address the impact of mediated communication on society. This includes an overview of scholars belonging to the Toronto School of Communication, who have provided an in-depth analysis of the impact of early forms of mediated communication on society. Then the chapter reviews modern trends in mediated communication and discusses the complex ways in which these trends affect interaction and community. While most discourse focuses on the benefits of mediated communication, scholars have also warned about the potential negative effects on people's social life. Do technologies allow us to maintain strong, rewarding, and supporting relationships? How do communication technologies overcome constraints of time and space? The chapter then focuses on how social media have redefined our notion of friendship and examines the implications of these changes for social networking. This leads to an analysis of romance on the Internet, investigating how people form and terminate romantic relations using social media. Finally, the chapter ends by exploring the concept of virtual mourning and how people renegotiate the meaning of death. To examine this new concept, we review the unprecedented online response to the death of Michael Jackson (the "King of Pop") and the online community that formed to support Canadian Eva Markvoort during her struggles with cystic fibrosis.

Early Beginnings of Mediated Communication

Abstract paintings found in Namibia that predominantly depict animals are the earliest forms of expression and date from 26,000 to 28,000 years ago. While these early paintings are simple and primarily symbolic in nature, recent anthropological evidence suggests that they are the result of the single most important human trait. From the early beginnings, humans seem to have had a fundamental need to express themselves (Crowley & Heyer, 2011). When the first humans arrived in Europe, the climate was harsh and they had to share their living space with the Neanderthals, who had populated the European continent much earlier. Neanderthals mysteriously disappeared between 250,000 and 30,000 BCE, probably crowded out by humans. Recent anthropological evidence suggests that humans and Neanderthals were surprisingly similar in many ways—skull size, dexterity, strength, etc. One distinguishing feature that scholars think gave humans an evolutionary advantage over Neanderthals was the extent to which various human social groups communicated. This perhaps allowed the human species to coordinate resources better, to learn about various hunting techniques, and to maintain a sense of social cohesion.[1]

With the introduction of writing, mediated communication became more elaborate. While early **hieroglyphs** are depictions of actual objects, later hieroglyphs represent sounds, providing a more flexible system of representation (Crowley & Heyer, 2011). In the case of Egyptian tombs, written texts were meant to be permanent and hence were often inscribed in stone. As a result, **epigraphy**—the study of the history and social circumstances of written information preserved on hard materials—emerged to provide unique insights into the social structure of these ancient civilizations. Ancient civilizations also used writing to preserve records, as in the case of banking archives from ancient Crete. A common use of writing was to document religious rituals and prayers. Moreover, the **impermanent record** is a category of ancient written inscription that was not intended to be preserved, but the ones that have been saved (either preserved in rock or found in tombs) can offer valuable insights into the social structure of ancient cultures.

The Toronto School of Communication

The study of how mediated communication impacts society has a long tradition, starting with the early investigations from the **Toronto School of Communication,** which was a loosely connected network of scholars who resided at the University of Toronto. Key intellectuals of the school include Eric A. Havelock, Walter J. Ong, Harold A. Innis, and Marshall McLuhan. In this section, we briefly discuss some of the most central theories and ideas put forward by these scholars.

Havelock went back in history and investigated the first technologies of communication. Havelock's theorizing was instrumental in stressing the political, social, and cultural changes brought about by the move from **oral societies** to **literate societies** (Crowley & Heyer, 2011). In oral societies, memory is used as the primary means for remembering events, understandings, and knowledge. In literate societies, on the other hand, the written word becomes the authoritative means of recording information (Havelock, 1963). For Havelock, the move to literacy in Greek culture led to two fundamental shifts:

1. **Content of thought**: Writing changed consciousness and thought processes. Havelock sees Plato as the first to express the change in thought. In oral societies, the focus is on action, either internally or externally motivated, and on the thoughts accompanying these actions—sorrow, happiness, or content. By contrast, writing brings with it more abstract and propositional content (Halverson, 1992).

2. **Organization of thought**: This is perhaps the most fundamental shift observed from oral to literate society and consists of changes in how main ideas and subordinate ideas are linked to one another. Havelock argues that in oral society, ideas are connected in an associative manner while in literate society they are organized to put forth a main idea to which all subordinate ideas are logically linked.

Havelock influenced many scholars in the fields of communication, sociology, and literary studies. His work, however, has received strong criticism from academics because of its lack of rigour, limited empirical evidence, and inappropriate methodology (Halverson, 1992). In addition, critics claim that Havelock did not have a full understanding of the ancient texts on which his main arguments were based.

Ong (1991) continued in Havelock's footsteps, investigating the social changes resulting from electronic media, such as television, the telephone, and radio. His key intellectual contribution was to move away from a simple dichotomy between the written and the oral. He argues that different notions of orality could exist, some of which directly draw on print culture. Ong distinguishes between **primary** and **secondary orality** to understand communication. He argues that second orality represents a form of **post-typography** because in it oral communication is dominant over the written word. For him, second orality describes a second verbal era that integrates elements from both oral and literate societies. This form "has striking resemblance to the old in its participatory mystique, its fostering of a communal sense, its concentration on the present moment, and even its use of formulas" (Ong, 1991, pp. 133–4). Ong makes it clear that despite the resemblance between oral societies and secondary orality, the latter is more deliberate and self-conscious, directly linked to writing and print.

An equivalent line of thinking emerged in the writings of Innis and McLuhan. Innis (1951) emphasized the economic, structural, and social changes resulting from literacy. Central to his theorizing is the concept of **media bias**, which refers to the transforming power of media in human affairs. He distinguishes between forms of communication that have a **time bias** versus those that have a space bias. Time bias describes oral societies in which there is an emphasis on community building. In these societies, change occurs slowly because the only transfer of culture, information, and knowledge is through oral means. By contrast, literate societies have a **space bias** that favours imperialism and commerce because writing allows for the easy dissemination of ideas and information over vast distances. Innis describes Ancient Greece as an example of an oral society with a time bias that supports community and is averse to change. By contrast, Innis points out that the Roman Empire had a space bias because the Romans focused on expanding their influence throughout Europe, Africa, and the Middle East. For this purpose, written documents were important as they allowed for messages, new rulings, and other types of information to be conveyed throughout the empire. Other examples of space-binding media include print, radio, television, and digital media because they reach a wide range of people and overcome space constraints.

McLuhan is probably the most renowned academic from the Toronto School of Communication. He examined the impact of print on culture, but went further with his focus on electronic media (McLuhan, 1964; McLuhan & Powers, 1989). McLuhan's well-known aphorism **the medium is the message** underlines the need to examine the impact of media on people instead of focusing only on content. Most of the previous research had examined the content that media present to people and how individuals, groups, and societies make sense of that content (see the discussion in Chapter 2). McLuhan (1964) advocated for the relevance of the characteristics of media because he was convinced that this had a much more pervasive and all-encompassing influence on society than content alone. He studied the impact of television on North American society and concluded that television not only compressed time and space but also radically transformed the individual's capacity to process information. McLuhan is often seen as a visionary because many of his aphorisms, such as the **global village**, continue to resonate with the way the Internet is bringing people together from across the globe, for example, in virtual communities.

The Toronto School of Communication provided many early insights into how media affect society. Nonetheless, the group only scratched the surface in terms of the many interrelationships that exist between prevalent modes of communication and the structuring of society. In the next sections, we examine first how the telephone slowly became adopted in North America and then examine the effects of the Internet on forming and maintaining social relationships.

North America Calling: The Impact of the Telephone on Social Relationships

A proliferation of innovations occurred in a number of different fields throughout the nineteenth century, including transportation, electric lighting, the radio, and the telephone. These advances were part of the changes occurring as a result of the Industrial Revolution. Prior to these advances, communication technology had been rather rudimentary. For example, flags, beacons, and smoke signals—referred to as **optical telegraphs**—were common forms of distant communication that predated the electrical transmission of messages. Then in 1794, Claude Chappe developed in France a new system that was effective in transmitting messages via towers spaced 8 to 15 kilometres apart, stretching a total of 4800 kilometres. The **Chappe communication system** was based on a semaphore relay station, which allows conveying messages through visual information. The semaphore telegraph relied on "a large horizontal beam, called a regulator, with two smaller wings, called indicators, mounted at the ends, seemingly mimicking a person with wide-outstretched arms, holding a signal flag in each hand" (Holzmann & Pehrson, 1995, p. 59). How the indicators and the regulator are positioned could vary, allowing for the system to transmit up to hundreds of symbols. To a large extent this system is similar to Morse code, which transmits messages with a series of on-off tones, with the key difference being that the Chappe communication system relies on the transmission of visual information. After several iterations, for instance, this system allowed Napoleon Bonaparte to communicate effectively with his troops because the system was very useful in transmitting messages quickly over long distances, resulting in its widespread adoption in other parts of Europe.

Innovation continued in the realm of telecommunications, leading to the invention of the telegraph, a device that for a long time was the main means of transmitting printed information by wire or radio wave over long distances. The telegraph was followed by the introduction of the telephone, which slowly, albeit drastically, changed the way people communicated with one another.

The telephone has undergone many changes since Alexander Graham Bell developed the first prototype (Klemens, 2010). The initial model only allowed for the transmission of sounds: speech could not be discerned. Around 1910, the **party line** (or shared line) was introduced, making the telephone more user-friendly and affordable. **Automatic dialing** was introduced in the 1920s, resulting in a complete shift in how telephones were used because there was no longer the need for operators. While telephones have become normalized in society, the period of early adoption was a difficult transition.

Claude S. Fischer (1992b) writes about the resistance to the telephone on the part of diverse groups and the ways in which the telephone fundamentally changed American society. Moreover, Fischer writes that "[a]s much as people adapt their lives to the changed circumstances created by a new technology, they also adapt that technology to their lives" (1992b, p. 5). Hence, the impact of the telephone was not radical; instead, people gradually came to rely on the telephone over time as the technology became embedded in existing norms and practices, eventually functioning as an additional form of communication.

One social change that the telephone brought about was the rewriting of boundaries set out by class, race, and status (Marvin, 1988). The telephone created a new form of accessibility, where people who were previously inaccessible, such as public figures, could suddenly be reached directly and instantly. With this new accessibility "[n]ot even the famous, those who are widely known but personally remote, were exempt from the reorganization of social geography that made socially distant persons seem accessible and familiar" (p. 66). As the technology has evolved, however, people have gained more control over how they can be reached and by whom. Quan-Haase and Collins (2008) investigated accessibility via instant messaging (IM) and concluded that people have more control over their social accessibility in instant messaging than with the telephone because the user can determine when to log on the system, when to log out, who to ignore, and with whom to communicate. Megan, one of the participants in the study, explains how it works: "Usually [IM] goes right away to 'away.' Because it keeps the people who are just bored and need somebody to talk to from talking to me, 'cause then they know that I'm away. But if it's something important, then they'll talk anyway." This shows how features in the technology provide users with more control over their social accessibility.

The telephone also shifted geographical boundaries in that it opened up previously sheltered social realms, such as the home: "Home was the protected place, carefully shielded from the world and its dangerous influences. New communications technologies were suspect precisely to the extent that they lessened the family's control over what was admitted within its walls" (Marvin, 1988, p. 76). Despite people's initial apprehension, the telephone became widely adopted in Western societies and has had widespread social consequences. Nonetheless, the telephone was just the beginning of the close link between telecommunications and society. The Internet as a medium would diffuse much more rapidly, and its effects would be more profound and all-encompassing.

Penetration of Mediated Communication: The Impact of the Internet on Social Relationships

We have seen a dramatic increase in Internet use since the 1990s, affecting the way people live, work, and play in the developed world. For a large proportion of Internet users, Internet access is a daily activity. Communication via the Internet is referred to as **computer-mediated communication** or **CMC**. Generally, we make a distinction between **asynchronous** and **synchronous forms of communication**. Asynchronous communication refers to exchanges where communication partners do not need to be present at the same time, such as email or wall posts on Facebook. Synchronous communication refers to communication that occurs in real time. The most well-known form of synchronous communication is instant messaging, including Skype, Windows Live Messenger, and AOL AIM. We can also distinguish a set of applications that fall under the rubric of **social media** and that have been defined as "a group of Internet-based applications that build on the ideological and technological foundations of Web 2.0, and that allow the creation and exchange of User Generated Content" (Kaplan & Haenlein, 2010, p. 61). Social media tend to be more holistic in that they integrate elements of asynchronous and synchronous forms of communication (Hogan & Quan-Haase, 2010). Examples include **social network sites (SNS)**, **wikis**, and **blogs**.

The Internet as a form of communication has come to occupy a predominant role in our society. Figure 9.1 shows how penetration rates have changed in the United States from 2000 to 2009. While there are some minor peaks in the data, the trend is for an increase in Internet use among all age groups. Adults 65 and older are still the slowest to adopt the Internet as a communication tool: only 38 per cent in this age group used the Internet in 2009. This group of users is often referred to as **digital immigrants** because they came to the Internet late in life and have been generally slow in adopting new forms of communication (Prensky, 2001). In the 12- to 17-year-old and 18- to 29-year-old age groups, 93 per cent are users, which represents an increase of almost 20 per cent since 2000. This group is usually referred to as **digital natives** because they have grown up with the Internet and are unfamiliar with a pre-Internet time (Palfrey & Gasser, 2008; Prensky, 2001). In addition, the data show that 63 per cent of teens go online every day, of which 36 per cent go online several times a day and 27 per cent go online about once a day.

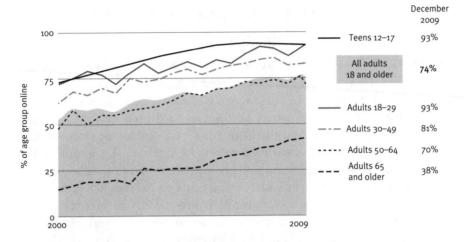

FIGURE 9.1 CHANGE IN INTERNET USE BY AGE, 2000–2009
Source: Lenhart, A., Purcell, K., Smith, A., & Zickuhr, K. (2010). *Social media and mobile Internet use among teens and young adults.* Washington, DC: Pew Research Center, p. 6.

Since 2005, we have observed a shift in how people use media for communication. Email and instant messaging—even though they continue to be widely used—have been largely displaced by social media (Quan-Haase & Young, 2010). In Canada, for example, about half of the population spends no less than one hour a day on **Facebook** (Moretti, 2010; see Figure 6.2 on p. 116). Other research shows that, in the United States, 73 per cent of online teens (ages 12–17) and 72 per cent of young adults (ages 18–29) use SNSs (Lenhart, Purcell, Smith, & Zickuhr, 2010). However, only 40 per cent of US adults 30 and older use SNSs (see Figure 9.2). Similarly, in Canada the Facebook population above 35 years of age accounts for only 27 per cent of the entire user group (Moretti, 2010), which suggests that the older populations, while not completely absent in social media, are lagging behind when it comes to its adoption.

While a range of providers, such as hi5 and Friendster, emerged in the early days of SNSs, Facebook and MySpace dominated the market in 2011. Among Americans 18 and older, Facebook was the most popular SNS with 73 per cent of adult users having a Facebook account, 48 per cent having a MySpace account, and 14 per cent having a LinkedIn account (Lenhart et al., 2010). And Americans are not the only ones to embrace Facebook as their SNS of choice: Canadians have embraced the technology in large numbers and are the greatest users of Facebook worldwide. In Canada, 77 per cent of 13- to 17-year-olds and 82 per cent of 18- to 24-year-olds use Facebook more than email (Moretti, 2010). In the summer of 2011, Google+ was launched as an alternative social networking site that allowed users to add friends, acquaintances, and family within distinct *circles.*

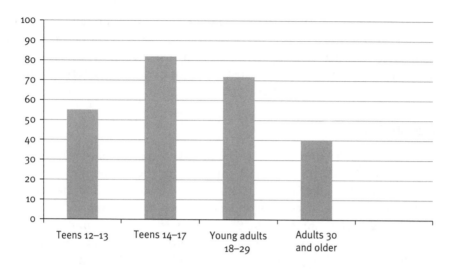

FIGURE 9.2 USE OF SOCIAL NETWORKING SITES IN THE UNITED STATES
Source: Lenhart, A., Purcell, K., Smith, A. & Zickuhr, K. (2010). Social Media and Young Adults in The Pew Internet and American Life Project, www.pewinternet.org/Reports/2010/Social-Media-and-Young-Adults.aspx.

Social media have revolutionized the communication landscape, becoming an integral part of how we communicate. We examine next how these tools have changed how we understand friendship, how we form and maintain relations with friends and romantic partners, and how we present ourselves online.

How Has Technology Affected Our Relationships?

We don't usually tend to think of social phenomena, such as friendship, family ties, and romance, in terms of technology. Technology, however, does play a significant role in how we build and maintain social relations. In the next three sections, we describe how digital tools have impacted the fabric of our relationships, fundamentally transforming how people relate to one another. We start by investigating how our understanding of friendship has changed since the introduction of social media tools, such as Facebook and Twitter. The next section deals with romance online, a topic of great relevance with the increase in dating sites, such as eHarmony, PlentyofFish, and Lavalife. In particular, we focus on what has been termed the breakup 2.0, the understanding of how digital tools impact the dissolution of relationships. Tools leave digital traces behind that can make it difficult to move on and leave previous relationships behind. In the final section, we delve into the topic of virtual mourning. Even though our digital traces may remain in perpetuity online, our material counterparts are much more vulnerable. As loved ones—family, friends, colleagues, co-workers—pass away, we need to come to terms with this loss, and the online environment can provide an important community and source of comfort in these difficult times. We discuss recent research on virtual mourning—the grieving experienced online as a result of the death of a loved one—and virtual stranger loss—the online activity that results from the death of a celebrity or public figure.

Friendship Online

What does friendship mean on social media? When we have accumulated hundreds of friends on Facebook, this changes the meaning of the word. Traditionally, **friendship** reflects an informal association that has no distinct boundaries. Unlike work and family ties, friendships are voluntary in nature, and the content of the friendship relation is determined, negotiated, and established by the two parties involved and can change over time (Fischer, 1982). There is no single meaning of *friendship* as it is culturally determined and based on social norms. Nonetheless, there is an understanding that friendship relations are strong, based on trust, and provide emotional, economic, and social support (Granovetter, 1973).[2]

On SNSs, people send others friendship requests to connect them to their profile, a process called **friending** (boyd, 2006). An assumption people made in the past was that the list of people on Facebook directly reflected a person's **social network** and hence was a reflection of a person's circle of close friends. This is not the case. Research has shown that users link their SNS profile to a fairly disparate mix of people instead of it being a direct reflection of one's network of close friends only (boyd, 2006). Users often add people they have met only once or who are friends of friends.

The reason friendship acquires a different meaning in social media becomes evident when we focus on the features of the technology; SNSs introduce new constraints and possibilities, creating new norms around friendship formation and maintenance (Baym, 2010). Haythornthwaite (2005) introduces the term **latent ties** to describe the process whereby "adding any network-based means of communication—whether a new IRC channel, a social support group, a Webboard or email listserv—lays the groundwork for connectivity between formerly unconnected others" (p. 136). Latent ties refer to relationships that have yet to be formally developed but have had the foundation set for friendship formation through the networking power of technology. Features of communication technologies have a direct effect, therefore, on how friendships form and develop.

To provide a more systematic framework of the effect technology has on friendship, danah boyd (2006)—in her research based on users of **Friendster** and **MySpace**—identified four different features of SNSs that affect friendship formation:

1. **Persistence**: Data posted on profiles and walls of SNSs remain indefinitely archived on these sites and can be retrieved in the future.

2. **Searchability**: This is a key feature of digital data because it facilitates finding information on people.

3. **Replicability**: This feature refers to capabilities provided online to reproduce content (text, pictures, and videos) and insert it in other contexts. Users can post content effortlessly from one medium to another or from one conversation into another.

4. **Invisible audience**: Determining the identity of a recipient or reader of a post is not easy, and thus tailoring messages toward specific audiences is also difficult.

These four characteristics of online media are important in terms of understanding the meaning of friendship online. On Facebook, for example, people can see who someone else's friends are; this puts pressure on peers to have a similar number of friends, to have similar kinds of friends (status, gender, diversity), or to have specific individuals as friends as well. This transparency, which was lacking before the introduction of SNSs, can create tension, change social norms around what friendship means, and create expectations around friendship formation. That same transparency can also help activate latent ties, which in turn can provide useful and new information, social support in difficult times, and companionship when away from home (Quan-Haase & Young, 2010).

Another important feature of online environments is that users need to explicitly declare friendships. Boyd argues that, through the activities of friending, "people **write community into being** in social network sites" (2006, p. 1). By making choices about who to add to the list of friends and who to exclude, people are drawing the boundaries of what their community looks like. Displaying their social connections provides a means to define who they are because "[a] public display of connections can be viewed as a signal of the reliability of one's identity claims (Donath & boyd, 2004, p. 73). These public expressions of friendship provide rich material for researchers to obtain a better understanding of social phenomena as they unfold online.

To what extent does the Internet facilitate the formation of new relationships? While most users of SNSs have large networks (130 on average), these relationships are rarely initiated online. While early research suggested that computer-mediated communication would facilitate the formation of new social ties, accruing evidence shows that such communication is used primarily for maintaining existing social ties (Ellison, Steinfield, & Lampe, 2007; Katz & Rice, 2002). New relationships would mean meeting total strangers online and getting to know them well through the Internet. While new relationships do form online, as in the example of Eva Markvoort who we will discuss later in this chapter, the majority of online exchanges occur between friends, family, co-workers, and classmates.

One of the key issues that persists with friendship formations on the Internet involves the concept of identity (Baym, 2010). Most users present an identity online that largely conforms to or is truthful to who they are in the offline context (Back, Stopfer, Vazire, Gaddis, Schmukle, Egloff, et al., 2010; Baym, 2010). At the same time, Mehdizadeh (2010) argues that on social media individuals have some freedom to customize their profile and promote themselves with attractive photographs, fun status updates, and postings of their achievements. Most people we "friend" on SNSs are people we have met before in person or people who are friends of friends. In fact, for the large part, SNSs are a representation of our offline social networks, which limits the extent of self-idealization.

Romance Online

Understanding social media also includes getting a sense of its effect on romantic relationships. People use social media, first of all, to screen potential romantic partners by searching Google and visiting their social network profile. In addition to providing verifying information on potential romantic partners, social media can also move the relationship forward more quickly, as text messaging, posts on walls, and Twitter tweets can fill the communication gaps that might otherwise occur in between in-person meetings (McGinn, 2011).

Tufekci (2007) found that about 60 per cent of university students indicated their romantic status and their sexual orientation on their social network profile. While students indicate that they are concerned about strangers accessing their profile, they also see such availability as advantageous: potential romantic partners can learn about each others' interests, meet friends of friends, and find out about a person's unique personality through wall posts, reports of past activities, and pictures. In part, the social network profile provides a glimpse into someone else's past and present life. Using social media to get to know someone is fun and exciting because individuals can obtain new and unexpected pieces of information about the other person. However, recent studies on **breakup 2.0**, which we discuss in more detail in Box 9.1, show that those same benefits of social media can also be problematic during the dissolution of a romantic relationship.

Box 9.1 ⁘ Breakup 2.0

Ending romantic relationships has never been easy. Gershon (2010) was intrigued by how new media has affected the formation and dissolution of romantic relationships. In her investigation, she found that young people are confused about the role that mediated communication should play in the breakup.

Gershon based her study on 72 interviews and 472 questionnaires primarily with university students. She discovered that people believe that the medium chosen to communicate the breakup is pivotal. Some media are evaluated as more serious than others and hence are more suited to end a relationship. For most participants in the study, face-to-face communication was the medium of choice for breaking up with someone, and they considered using other media to be highly inappropriate. Texting, for instance, was considered the least appropriate medium for ending a relationship.

In addition, Gershon identified problematic breakup practices unique to social media: changing one's relationship status on Facebook, re-reading and overanalyzing emails and wall posts, stalking online, and deciding to "defriend" an ex from Facebook. Kate Miller-Heidke, in her tongue-in-cheek "Facebook Song," hilariously captures the essence of the Facebook breakup. She tells the story of how difficult it was to forget her ex: all the pain, agony, and energy that it took her to get over the relationship. Then out of the blue she gets a friendship request from her ex on Facebook. In her lyrics, she expresses her fury over the invasion of her privacy and over the thought of having to deal with his constant virtual presence. To listen to a version of Miller-Heidke's song, go to www.youtube.com/watch ?v=QeTPNjPiyJM&feature=related.

Box 9.1 shows how social media have made the process of breaking up more difficult. As a relationship unfolds, it leaves a digital trace on Facebook (Hogan & Quan-Haase, 2010), which later triggers memories of the romance even after its dissolution. José van Dijck (2007) in her book *Mediated Memories in the Digital Age* points toward the existing relationship between material culture, technology, and memory, arguing that objects mediate memory and represent a person's identity in a specific moment in time. In the same way in which people fill shoe boxes with personal items, social media become repositories of our past relationships. These digital traces can make it difficult to move on and leave previous relationships behind.

In the context of romantic relationships, the use of media follows pre-established social norms, often established over time within specific user groups. Gershon (2010) identifies three key concepts to explain how people adopt social media in their daily communication practices:

1. The term **media ideologies** describes the beliefs users form about computer-mediated communication and the perceived meaning assigned to the various forms of communication. *Media ideologies* is based on the concept of **remediation,** which describes how new media refashion prior media forms (Bolter & Grusin, 1999). For example, many university students perceive email as a formal medium because it resembles a letter.

2. **Idioms of practice** refer to how groups of people agree to use different media. The term describes the social norms that evolve around media use through conversations, the sharing of stories, and advice from others about appropriate media use.

3. **Second order communication** conveys the formality and nature of a medium. For instance, texting (or text messaging) is usually perceived as informal because of its brevity and lack of interactivity.

These conventions about how and when to use a specific medium are relevant social norms that help guide social behaviour. The social norms that have emerged on the Internet regarding communication and behaviour are referred to as **netiquette**. Sometimes the medium can be at odds with the message, as in the case of texting to break up with someone. Gershon (2010) argues that new media are not intrinsically formal or informal but that social groups establish norms about which media they see as appropriate for what kinds of messages and social circumstances. Breaking up over texting, again, is perceived as completely inappropriate because texting is an informal medium that provides little possibility for feedback and engagement.

Mourning Online

Part of our culture is to express our sympathies to friends and family who have lost a loved one. In these tragic situations, family and community come together to celebrate the lost life and to provide each other with emotional support and comfort. We are observing a change in how mourning occurs as death becomes a part of the Internet (Hogan & Quan-Haase, 2010).

Even though virtual communities in which members build strong social connections have existed since the 1980s (Rheingold, 1993), these were in the minority back then (Castells, 2001). In early Net culture the emphasis was on exchanging information and collaborating, as the Web was the domain of academics, geeks, and the military. What we are currently witnessing is a move toward the **domestication** and **mainstreaming of the Web** as more people become users of social media and integrate social media into their daily lives (Hogan & Quan-Haase, 2010). That is, people are using the Internet not only as a source of information but as a means to socialize and to maintain their social relationships. A consequence of this shift has been the emergence of death as a salient factor online. As a result, the Internet becomes what Sofka (1997) has referred to as a **thanatechnology**, which consists of "technological mechanisms such as interactive videodiscs and computer programs that are used to access information or aid in learning about thanatology topics" (Sofka, 1997, p. 553) as well as a way to cope with the death of a family member, friend, or member of the community.

With millions of individuals creating profiles on social media sites, a unique Net culture will inevitably evolve to deal with loss. Within this new Net culture, we are observing two different kinds of mourning:

1. **Virtual mourning**: When a family member, friend, or classmate dies, people come together online to commemorate, mourn, sympathize, and provide emotional, social, and economic support.

2. **Virtual stranger loss**: This constitutes the online mourning of people who are strangers, that is, the mourning of people one has never met, either in person or online.

Virtual Mourning

Virtual mourning is becoming a common phenomenon. On SNSs, such as Facebook and MySpace, a unique link exists between an individual's profile and the person behind the profile. Jenny Sundén (2003) has argued that the profile represents the self as people **write themselves into being** through their online participation. Through all the forms of expression available online, users create a very personal portrait of themselves and display a vivid and nuanced **online persona**. Bernie Hogan (2010) has pointed out that while we tend to merge the SNS data with the person, all of this submitted content continues to exist outside of the self in a sort of virtual repository. As long as the user's profile continues to be updated through posts on walls, through pictures, and through status updates, the illusion persists that content and user are one entity. Hogan and Quan-Haase point out that "when the submitter dies, and the profile lives on, we can see this distinction all too clearly. Death has become the ultimate arbiter of this difference between the data that persists and the individual that does not" (2010, p. 311).

Carroll and Landry (2010) deal with the persistent online presence of deceased individuals, whom we knew personally, and how their MySpace and Facebook pages are transformed either literally or metaphorically into memorial pages and sites of grieving. In their study, the researchers identify four activities that take place on SNSs after loss.

The first activity is the creation of a narrative of the dead person's life. People tell stories about the deceased to cope with the loss. This engagement with the person's life history is a collaborative process, where people often negotiate and contend the person's various and divergent portrayals.

The second activity is to include on a person's wall short, simple notes to publicly express feelings of loss and solidarity, which allows the community to share with others their feelings and join in the act of memorializing and coping with grieve—for example, "RIP" or "I miss you." Following is another example from Carroll and Landry:

i miss you adam, i wish you could come back but since you can't, i wanted to say that i love you very much, we all miss you, ♥courtney (posted 9 April 2007, http://profile.myspace.com/index.dfm?useaction=user. viewprofile&friendID=175255389)

A third activity is to praise the deceased and highlight his or her virtues, merits, and accomplishments. A fourth activity includes addressing the deceased directly by asking him or her for guidance, comfort, and understanding. This act shows how powerful profiles are in continuing to represent the deceased and in providing some form of comfort for those left behind. The engagement in these four activities brings the community closer together, and the profile of the deceased provides a central space for people near and far to come together, to share their memories, and to express their feelings of loss.

Virtual Stranger Loss

A second, and very different, form of mourning on the Internet is stranger loss mourning. Stranger loss mourning can occur in a variety of ways, but perhaps the most pervasive form occurs when you mourn the loss of someone you have never met. On the CBC radio show *Spark*, producer Daemon Fairless retells a story where he found himself mourning the death of someone with whom he had never communicated with or met. He was searching online for do-it-yourself solutions on how to build bikes, when he unexpectedly came across Sheldon Brown's website.

Homepage of Sheldon Brown's Bicycle Technical Info website.
Source: www.sheldonbrown.com.

On the site, Brown had posted extensive information about bikes and how to fix them, as well as information of a more personal nature, such as his passion for bikes and pictures of him biking together with his family. After Fairless finished building his bike, he was pleased with his new creation and decided to contact Brown to thank him. While looking for Brown's email address online, he stumbled upon his Wikipedia entry, only to find out that Brown had passed away. Fairless was shocked. He was even more surprised about his reaction to the news: "I read some guy's Web site, it was not like we hang out or did some amazing adventures together. I felt foolish; I let myself develop a relationship with what? A Web site? It struck me as ridiculous."[3] This demonstrates how people often underestimate the power of the Internet for forming social connections even with those we have never met. Nonetheless, relationships online are as real as any other relationship we form as we illustrate in Box 9.2 with the story of Canadian Eva Markvoort, who built an extensive and loyal community around her.

How can we understand why people feel so strongly for others they have only met online? How does the medium change the dynamics of friendship, community, and social support? Without the Internet, the struggle of someone such as Markvoort (see Box 9.2) would have taken place privately. Traditional media, such as radio or television, would have perhaps provided a short documentary on her journey, which would have been broadcast to the audience without allowing any significant amount of intimate engagement with the person and the material. In the context of traditional media, it would have been difficult to directly react to a story like Markvoort's and to get in touch with her.

Box 9.2 ❖ 65 _ RedRoses: The Community around Eva Markvoort

Eva Markvoort, a 23-year-old Canadian from British Columbia, was diagnosed with cystic fibrosis (CF), a potentially deadly disease. While undergoing treatment and waiting for a double lung transplant, Markvoort created a **blog** on **LiveJournal** about her experience under the name 65_RedRoses. The blog started on 15 July 2006:

> I need somewhere to vent. To let go and not always be motivating and inspiring. I need somewhere to let my fears go unleashed . . . I'm hoping to find someone else who knows what it's like to climb the stairs and not be able to breathe and hate the fact that you know your body needs a couple of weeks in the hospital. Someone else who is sick and tired of being sick and tired. Who appreciates the days when skipping and running is possible because those days are farer and fewer between.

Eva Markvoort as 65_RedRoses.
Source: Photo by Cyrus McEachern.

The blog presented Markvoort with an opportunity to reflect on her own struggle with CF and to communicate to a world out there on the Web about her journey. She probably never imagined the impact she would have on people's lives through her blog. Her vibrant personality and the tragic nature of her story created an outpouring of support, friendship, love, and understanding. As Markvoort's health went through up and downs, her online community provided her with comfort, reassurance, and strength. Through her blog entries, Eva touched the lives of many people who either had previous experiences with CF or who felt moved by her story and connected to her. She also got to closely know some of her online followers. Two girls whom she connected with were also suffering from CF, and by sharing their day-to-day experiences with the illness, the three became close friends. The community that formed around Markvoort was very powerful because it was based on a shared experience and it became for her a vital source of emotional support.

A key distinguishing feature of new media is **interactivity**, which describes users' ability to provide content in response to a source or communication partner (Ha & James, 1998). Hence, the Internet has removed a layer between the sender of a message and the receiver, allowing for a direct and high level of involvement from a larger audience. This provides audience members more control over content and its use than does traditional media. Many of Markvoort's blog entries, for example, contain comments, where readers left messages for her to read. In addition, she was able to communicate with people she met online via IM services (Skype), which allow communication partners to talk to each other in real time as well as to see each other.

With the domestication and mainstreaming of the Internet, loss, illness, and suffering have also inevitably become a part of being online. The Internet serves as a space for friends, family, and the larger community to come together and express their feelings and thoughts. It is a unique platform that overcomes barriers of time and space. Similarly, Eva Markvoort's story shows how the Internet allows those with similar life experiences to connect and provide each other with friendship and support.

A completely different kind of mourning was expressed online with the death of the "King of Pop," Michael Jackson, on 25 June 2009. When people found out about his death, they went online to confirm the news, to find out more information surrounding the events, to look up biographical data and pictures, and to read about the reactions of the Jackson family to the event. This move toward the Internet, and toward social media in particular, was unprecedented, leading to the slowing down of the Internet as a whole and the crashing of numerous websites due to the increased Web traffic.

The kind of mourning that is expressed when celebrities die is referred to as **para-social grieving** (Sanderson & Cheong, 2010), which refers to the loss of an individual whom the mourner did not know personally. The sense of relationship with the celebrity is based on repeated exposure through the mass media, a relationship referred to as **para-social interaction** because it does not include in-person interaction. Horton and Wohl (1956) described this phenomenon as first occurring on the radio, where listeners would form very strong and intimate bonds with individuals they had never met, but only knew from the radio shows. While these bonds are perceived as being very personal and real, they are in fact mediated by technology and are one-sided.

A study of the development of para-social grieving after Jackson's death on Facebook, TMZ.com, and Twitter showed that all the stages of grief identified in previous research apply to the online environment: acceptance, denial, anger, bargaining, and finally depression (Sanderson & Cheong, 2010). People first demonstrated shock after finding out that the singer had died. This was followed by disbelief, which often led people to seek out more information about the events on celebrity gossip websites or news sites. The denial phase was followed by anger and bargaining, resulting from not understanding how Jackson could have died at such a young age, and in light of his preparation for his new world tour. The final stage was feelings of depression and sadness resulting from the realization that Jackson was indeed gone.

In addition, the results show that participants in social media also used religious discourse to make sense of the loss of Jackson. Table 9.1 shows the most relevant themes that emerged, definitions of those themes, and examples as they appeared on social media sites. For instance, Theme I is labelled "worship" and encompasses posts with religious connotations that indicate worship of Michael Jackson. In these worships, often the words *King* and *Prince* were used to show the strong devotion toward the performer.

TABLE 9.1 **The Para-Social Grieving Process after the Death of Celebrity Michael Jackson**

Theme I

Label	**Worship**
Definition	Postings that invoke religious language reflecting worship of Michael Jackson.
Indicators	Using affectionate and/or intimate language with appellations, such as "King" and "Prince."
Differentiation	Postings that do not use appellations such as "King" and "Prince"
Example	"I am feeling a lot for the death of our King of Pop Michael Jackson." (TMI–271)

Theme 2

Label	**Coping**
Definition	Postings that use religious language to position Michael Jackson being in heaven or being in a "better place."
Indicators	Describing Jackson as being with a Divine Being, making references to Jackson being in heaven, and/or in a "better place."
Differentiation	Postings that do not designate Jackson as being in heaven, in a "better place," or with a "Divine Being."
Example	"Hope you're doing well in Heaven!!!" (FBI-50)

Theme 3

Label	**Empathy**
Definition	Posting that using religious language directed to helping others, particularly Jackson's family cope with his death.
Indicators	Invoking a Divine Being's blessing on others, or stating that they were praying for others to heal from this loss.
Differentiation	Postings that do communicate that one is praying for others or requesting a Divine Being's blessing to attend them as they cope with Jackson's death.
Example	"God bless the Jackson family in this time of grief." (FBI-65)

Theme 4

Label	**Condemnation**
Definition	Postings that use religious discourse to castigate Jackson and denote their contempt for him.
Indicators	Referring to Jackson as being in hell, or being glad that Jackson had died.
Differentiation	Postings that do use religious language to glorify Jackson and promote positive aspects of his afterlife destination.
Example	"Burn in Hell Michael!!!!!!" (TMI – 344)

Theme 5

Label	**Proselytizing**
Definition	Postings that using direct people to web content they have created to honor Jackson, to websites centered on Jackson, or to public gatherings to mourn Jackson.
Indicators	Requesting others to click on hyperlinks that direct them to content created material on the web that memorialized Jackson, websites about Jackson, or public gatherings to mourn Jackson.
Differentiation	Postings that do use hyperlinks to direct people to other websites honoring Jackson or announcing public gatherings mourning Jackson.
Example	"It's Michael Monday, let's celebrate Michael's voice, music, and songwriting talent, share his music with your followers." (TWI-22)

Source: Sanderson and Cheong (2010). 'Tweeting prayers and communicating grief over Michael Jackson online'. *Bulletin of Science, Technology and Society*, 30(5), p. 334.

Often, the online mourning of a celebrity is also accompanied by a flurry of activity on **Wikipedia**. After Michael Jackson's death, his entry became contested territory as people started adding contradictory information. Some fans protected the entry from critical, and what they felt were unfounded, remarks against the pop star, and they continued to emphasize Jackson's legacy in numerous music genres, on the music video scene, and in his acts of generosity. By contrast, an opposing camp formed on Wikipedia that emphasized the allegations of child molestation and his often eccentric behaviours. In the case of the death of a celebrity or public figure, Wikipedia becomes a central focus of activity, interaction, and contention, probably because people perceive Wikipedia as a place where public information on a person is displayed and archived, representing an important component of that person's online persona, even after death.

The virtual mourning that occurs on social media sites after the death of loved ones or high-profile individuals is a new phenomenon that was not possible in traditional media. Through formal obituaries, traditional media would play a key role in providing information about the death of celebrities and key public figures, but the audience would not have been able to express and share publicly its emotions, opinions, and reactions. Social media changes the dynamics, providing audiences on a global scale with new ways to express their mourning, to negotiate the meaning of a celebrity or public figure's death, and to undergo the phases of dealing with loss.

Conclusions

This chapter provides an overview of how technology mediates social relationships and community. The first systematic analyses were conducted by the Toronto School of Communication, whose members investigated the history and social circumstances of oral and written communication. The central contribution of the Toronto School of Communication was to demonstrate the importance of studying media in relation to society. Previous scholarship had neglected to understand the role that media play in shaping social, economic, and cultural factors. McLuhan, however, stated that to examine society, we need to understand that the medium is the message. With the introduction of the telegraph and the telephone, new opportunities emerged for mediating social relationships. Both tools greatly facilitated the transmission of information over long distances and across social spheres. For instance, the telephone led to the redrawing of pre-established geographical and social boundaries. The home, which was previously considered a sheltered and secluded social space, suddenly became open and easily accessible.

With the introduction of the Internet, there has been another shift in how people communicate and maintain social relationships at work, home, and school. The move toward the large-scale adoption of digital technologies helps users to stay connected 24/7 and anywhere. This ubiquity is possible through the widespread adoption of mobile technologies in combination with social media applications. And this move toward using social media for connectivity is certainly driven by young people. Researchers continue to grapple with questions about the social implications of this new move toward mediated connectivity. As more people flock toward digital technologies, we are witnessing a move toward the domestication and mainstreaming of the Web (Hogan & Quan-Haase, 2010). As part of this trend, new social phenomena emerge online around topics such as friendship, romance, and death. The chapter has discussed how "friending" is affected by a mix of social processes and technological affordances. Technology facilitates greater transparency about the nature of our social networks. This, however, creates new complexities in how we navigate our social spaces.

Questions for Critical Thought

1. How did the introduction of the telephone change the boundaries of the home in American society?

2. Discuss the relevance of age in current adoption patterns of SNSs. What key factors determine these trends?

3. Discuss the complex link between a person's online persona on a SNS, such as Facebook, and the self. To what extent is the online persona a true reflection of the self?

4. Relationships formed online are often dismissed as shallow and unimportant. Discuss what makes online relationships meaningful.

Suggestions for Further Reading

Baym, N.K. (2010). *Personal connections in the digital age.* Cambridge, UK: Polity Press. Addresses the debate about how information and communication technologies have transformed friendship, community, and family ties.

Boyd, D. (2006). Friends, friendsters, and top 8: Writing community into being on social network sites. *First Monday, 11.* Retrieved 15 December 2008, from http://firstmonday.org/htbin/cgiwrap/bin/ojs/index.php/

fm/article/view/1418/1336. This article addresses the changes that have occurred in the meaning of the word friendship as a result of social media.

Gershon, I. (2010). *The breakup 2.0: Disconnecting over new media.* Ithaca, NY: Cornell University Press. The book shows how the dissolution of romantic relationships has changed in the context of Web 2.0 technologies and in particular social media, such as Facebook and MySpace.

Quan-Haase, A. & Hogan, B. (2010). Persistence and change in social media. Two special issues of the *Bulletin of Science, Technology and Society, 30*(5 & 6), 309–15. This double special issue compiles 13 articles on current trends in social media.

Online Resources

Association of Internet Researchers (AOIR)
www.aoir.org/
AOIR is an international, interdisciplinary scholarly association comprising researchers with an interest in the advancement of Internet studies. The association hosts an open-access mailing list, has a website with extensive resources, and organizes an annual conference.

Pew Internet and American Life Project
www.pewinternet.org
The Pew is an American non-profit organization that conducts empirical research on how Americans use the Internet and the impact it has on their lives. The website is comprehensive, including reports, press releases, and statistics. Of particular relevance is their data on the connectedness of Millennials: www.pewinternet.org/Reports/2012/Hyperconnected-lives.aspx.

Interview with Rhonda McEwen on CBC
www.cbc.ca/freshair/episodes/2011/01/18/saturday-january-8/
This interview on *FreshAir* discusses Facebook as a tool to promote connectivity in society.

65°RedRoses: The Life of Eva Markvoort
www.cbc.ca/documentaries/passionateeyeshowcase/2009/65redroses/
This superb CBC documentary (with Force Four Entertainment) follows Eva Markvoort's brave struggle with cystic fibrosis (CF) and the mark she left on her online community. The site also includes a link to Markvoort's live blog and her memorial ceremony online.

10 The Surveillance Society

Learning Objectives

- ⊛ To define surveillance as a multi-faceted term with great relevance to the information society.
- ⊛ To discuss Foucault's analysis of power relations in society and to provide an overview of the concept and architecture of the Panopticon and its means of control.
- ⊛ To discuss how digital surveillance is changing our understanding of surveillance and how it is affecting individuals' privacy rights.
- ⊛ To present models of counter-surveillance as a means of personal resistance.

Introduction

The concept of surveillance has received considerable attention in society since the British philosopher Jeremy Bentham envisioned in the late eighteenth century the Panopticon as an ideal form of prison. The Panopticon is a building, a prison, and a tool, with its primary aim being to embody power and control. The goal of this chapter is to define the multi-faceted term surveillance by contrasting different perspectives available in the literature. The chapter then provides an overview of the concept and architecture of the Panopticon and its means of exerting control and imposing disciplinary action.

Originally, scholars thought of surveillance as a social process alone, and they viewed it as being only indirectly linked to technology. While institutions, and often buildings, were set up in such a way that allowed for surveillance, technology was not yet sufficiently developed to play a larger role in enforcing surveillance. Recent technological developments, however, have given surveillance a completely different meaning and have taken the controversy around surveillance to new heights. The new modes of surveillance show how technologies have changed not only the practices of surveillance but also its very nature. Indeed, to some extent these new modes of surveillance have greatly reduced individuals' privacy rights. For instance, surveillance as it occurs in social media has been amplified because in these environments everybody is watching everybody else. Finally, the chapter ends with a discussion of innovative methods of counter-surveillance that aim at increasing awareness of the pervasiveness of surveillance in our society and provide means for personal resistance.

Defining and Understanding Surveillance

The term **surveillance** originates in the French language and means "watching over." Observing the lives of other people and their behaviours, appearances, and social relationships is a naturally occurring social process. Innocuous and unobtrusive observing often occurs at coffee shops, when one sits at a table and leisurely watches passersby. Most of us have probably engaged in this kind of behaviour at some point or another. Similarly, since the inception of social network sites, many people have spent hours scanning through friends' photos, wall posts, and status updates. To describe this new and pervasive behaviour, users of social network sites have introduced the term **creeping**. As a result, the term surveillance has taken on a rather negative connotation. Most often, analysts link the term to the use of observation as a tool for exerting power and control, as often occurs in totalitarian regimes. Other related terms include *spying, supervision*, and *superintendence*. The *Oxford English Dictionary* (2011) provides a definition of surveillance: "Watch or guard kept over a person, etc., esp. over a suspected person, a prisoner, or the like; often, spying, supervision; less commonly, supervision for the purpose of direction or control, superintendence." Another related term with a negative connotation is **voyeurism**, which is the act of finding pleasure in secretly observing others engaged in private behaviours, such as undressing, sexual activity, etc.

Three Perspectives that Influence our Understanding of Surveillance

The terms associated with surveillance and their varied meanings show how academics have approached the topic from different directions. In addition to the various definitions that exist, Lyon and Zureik (1996) have identified three main perspectives that influence our understanding of what surveillance is: (1) **capitalism**, (2) **rationalization,** and (3) power. We discuss each perspective next to show its historical relevance to our current understanding of the term surveillance.

1. Capitalism

The first big shift in surveillance occurred during the time of industrialization. New forms of capitalist production in industrial-era England accompanied the desire to increase productivity, which resulted in new forms of control (Lyon & Zureik, 1996). Marx writes that "the work of directing, superintending and adjusting becomes one of the functions of capital, from the moment that the labour under the control of capital, becomes cooperative" (Marx, K., 1996, p. 217). Surveillance here takes on two roles: one as an internal component of production and another as a means of discipline (Foucault, 1995). The role of directing and supervising, originally occupied by controllers, was later substituted with machinery (see the discussion in

Chapter 6). As a result, the assembly line became an efficient means of control and surveillance that did not require any human intervention. Similarly, the introduction of clock-in cards and the design of offices as open cubicles were all part of a process of seamlessly embedding surveillance into work processes.

The study of surveillance from the capitalist viewpoint provides two advantages (Lyon & Zureik, 1996). First, it allows for an analysis based on **political economic theory**, where economic factors are the primary motivators for implementing surveillance practices, techniques, and tools. Second, this kind of analysis also "makes possible a critical stance in which systematic inequalities are exposed and a critique is made of major organizations and ideologies that perpetuate the system" (p. 6).

Criticism of the perspective stresses its rather narrow focus and points out two weaknesses (Lyon & Zureik, 1996). First, the idea that all surveillance is motivated by economic factors is limiting, considering that people engage in surveillance for social reasons as well. For instance, after a breakup ex-lovers often are obsessed with the behaviours of their former partners. Gary Marx (1996) describes how popular culture depicts this obsession—the song *Every Breath You Take* performed by the band The Police is a perfect example.[1] In the song, Sting writes about his obsession with his ex-wife after their separation, and also about his desire to watch every step she takes. The second weakness of the capitalist perspective is that the idea that surveillance is in and of itself negative reflects a simplified view of social processes. For example, users of social network sites provide detailed information about their whereabouts, their recent activities, their friends, and their current feelings. When friends, acquaintances, and even strangers spend time looking at this collage of information—a process referred to as creeping, as we mentioned earlier—they engage in surveillance. However, this is a form of surveillance that is welcomed by users (as we will discuss below). In general, then, the capitalist view of surveillance provides a good understanding of the economic meaning of surveillance in the context of work, but does not fully explain surveillance as it occurs in other social contexts.

2. Rationalization

The second big shift in surveillance is associated with changes in how society operates. Weber's concepts of *rationalization* and **bureaucratization** are closely linked with the concepts and practices of surveillance (see the discussion in Chapter 6). Rationalization describes a fundamental shift in the functioning of society: instead of behaviour relying on kinship ties, tradition, and informal affiliations, it now relies on rules based on **rational choice**. According to Weber (also see Kim, 2008), three factors that play a key role in this shift are as follows:

1. *Knowledge*: Knowledge is seen as the basis for rational choice. Rational choice requires an understanding of cause and effect as well as a weighing of various outcomes. Weber referred to this process as **intellectualization** because instead of relying on superstitious or mystical beliefs, decisions are based on modern scientific and technological knowledge.

2. *Growing impersonality*: Objectification occurs as part of rationalization, reflecting the Puritan's austere work and life ethic. Individuals, with their unique stories, are incorporated in a rational system, in which there is no consideration of personal concerns or matters.

3. *Enhanced control*: Most important for our current discussion, Weber saw rationalization as increasing control in social and work life. This increased control is a result of the Puritan ethic of self-discipline and self-control, which Weber referred to as **innerworldly asceticism**. Rationalization leads through institutionalization toward greater bureaucratic administration, legal formalism, and industrial capitalism.

The second term used in Weber's analysis is bureaucratization and refers to the establishment of efficient methods of organization that help governmental institutions and businesses fulfill their mandate. Often in these highly rationalized and bureaucratic workplaces, individual autonomy and freedom are, to some degree, replaced with manuals, functions, and pre-established roles.

Weber introduced the notion of the **iron cage** to provide an analysis of how nation-states, institutions, and modern organizations exercise power and control over their citizens, members, and workers, respectively. For Weber, there are two sides of the iron cage. On the one hand, rationalization increases an individual's freedom because it leads toward greater transparency in terms of how individuals can achieve their goals. Institutions outline the rules (the means) that individuals need to follow to achieve certain ends (the goals). This increases the predictability of social life in comparison to that existent in arbitrary authoritarian social systems. On the other hand, the iron cage seriously hampers human agency by narrowing down the possibilities and actions of individuals in a completely institutionalized system of rules (Kim, 2008). Weber writes that "No one knows who will live in this cage in the future . . ." as these individuals will be "[s]pecialists without spirit, sensualists without heart; this nullity imagines that it has attained a level of civilization never before achieved" (1920/2003, p. 182).

The Weberian perspective on surveillance has been criticized for its focus on technological change (Lyon & Zureik, 1996). Critics argue that this perspective sees technologies of rationalization, such as the assembly line, as the driving force of control. In addition, computers at work allow for further and more invasive forms of surveillance by documenting workers' activities. However, as Lyon and Zureik (1996) point out, the Weberian perspective provides an analysis of the social system itself and not directly of the means, either technological, social, or institutional, by which rationalization and bureaucratization are put in place. In the Weberian analysis, new technologies are not identified as the source for new forms of surveillance as some critics have incorrectly assumed.

3. Power

The third approach associated with surveillance is based on Foucault's (1995) work on disciplinary practices. For Foucault, power is an inherent part of all social relationships and social systems. As a result, social control becomes a central feature of modern societies. Foucault's analysis of prisons represents perhaps the most compelling and influential theoretical work on surveillance. To explore his ideas in more depth, the following section presents the concept of the Panopticon and how it relates to control in modern societies.

Foucault's Analysis of Power Relations in Society

To understand the social meaning of surveillance and the role technologies play in distributing and enforcing power, it is essential to examine first the concept of the **Panopticon**. In his analysis of how society treats, labels, and punishes crimes, Foucault traces the historical development of power relations in society. He argues that these modern forms of punishment do not only apply to prisons but rather have become the standard means of control for our entire society (Gutting, 2010). For Foucault, "**[d]iscipline** 'makes' individuals; it is the specific technique of a power that regards individuals both as objects and as instruments of its exercise" (Foucault, 1995, p. 170).

Foucault's analysis starts by tracing the history of punishment. In the Middle Ages punishment occurred as a public act of degradation, where the church and king penalized criminals in the public eye using torture and death by decapitation, burning, and starvation, which Foucault refers to as **punishment-as-spectacle** because the audience is as much part of the impartment of discipline as are the bearers of justice. In the eighteenth century, however, the methods of punishment to deal with disobedience changed radically. The population started questioning these shocking methods of domination, and the justice system replaced them with more "subtle" but equally powerful forms of subjugation. Foucault writes that "justice as a result no longer takes public responsibility for the violence that is bound up with its practice" (p. 9) but instead moves justice into the courtroom and the practice of punishment into a private, secluded space.

At the centre of Foucault's analysis of disciplinary action are three primary forms of control: the examination, normalizing judgment, and hierarchical observation (Foucault, 1995). The **examination** places individuals in a "field of documentation" (Gutting, 2010). Documents are records of various spheres of an individual's life and can serve as sources of power that help control an individual's behaviour. These records allow those in control to "formulate categories, averages, and norms that are in turn a basis for knowledge" (Gutting, 2010). The information gathered on the individual can then be used to examine that person as a single case in the context of socially established norms and values.

A central concern of modern disciplinary control is determining the extent to which people fall into specific categories and meet certain standards. Behaviour that does not meet the set standards would be considered **deviant behaviour** and could require disciplinary action. **Normalizing judgment** can establish whether an individual meets society's set standards or falls into the category of abnormal. Society has developed standards for many aspects of life. For instance, Canada has developed the Canadian National Standards for the Education of Children, which outlines the educational standards for core subjects for each grade and sets expectations as to what children should learn and when. Similar standards exist for medical practice, law, the environment, etc.[2] The introduction of standards marks a radical shift in how discipline is accomplished by establishing precise norms ("normalization"). In the older judicial system, single actions were judged in terms of whether or not they were permissible by law; this older system excluded a judgment about "normality" or "abnormality" (Gutting, 2010).

The most relevant of these forms for our present discussion is **hierarchical observation**, which describes how control over people can be achieved simply through surveillance. Observation occurs in a "network of gazes" that are laid out following a hierarchical structure, with data being conveyed from lower to higher levels. For Foucault (1995), implementing hierarchical observation requires a new form of construction in which buildings allow for "articulated" and "detailed" control. Control is articulated because the physical outline embodies the hierarchical nature of the observation. Moreover, such control is detailed in that the observer has a full view of the actions of those being observed. Basically, power relations are laid out in the design or shape of the architecture, as in the Panopticon, which we discuss next.

The Panopticon as a Means of Surveillance

The use of intricate systems of surveillance as a form of punishment started as early as 1780. Foucault (1995) carefully analyzes Bentham's Panopticon to show how in the design of buildings new forms of power can be embedded. The photo below shows how, in Bentham's design of a prison, a tower is located at the centre. The tower is surrounded by disconnected cells, and each cell houses a single individual, who is securely locked away. The arrangement is such that the supervisor, from a central point, can observe prisoners at all times without being noticed because all cells open toward the centre and the tower has visibility over all cells surrounding it. Hence, there is absolute transparency in terms of what the inmates are doing. By contrast, the lighting is set up in such a way that the inmates cannot see the interior of the tower. Moreover, they cannot even determine if anyone is in the tower at any given point in time. For Foucault, Bentham's model follows two principles: the **visibility of power** and the **unverifiability of power**. It is the combination of power being visible and unverifiable that is so effective in establishing discipline.

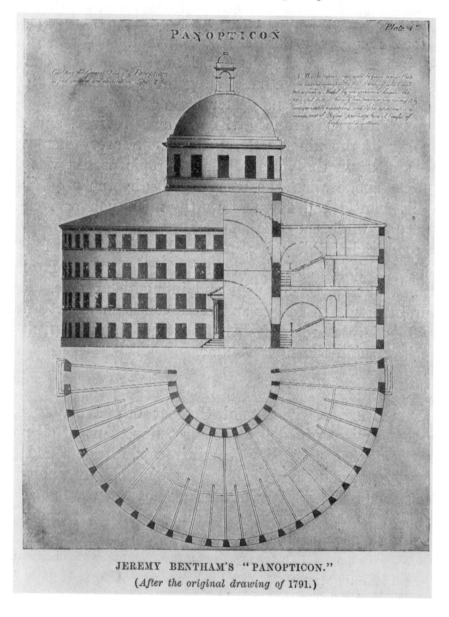

Jeremy Bentham's Panopticon (after the original drawing of 1791).
Source: © 2002 topham Picturepoint/GetStock.com.

The social consequence of this architecture is that prisoners cannot determine whether or not they are being watched. The Panopticon places the individual in a position where "[h]e is seen, but he does not see; he is the object of information, never a subject in communication" (p. 200). This uncertainty creates a new form of control as the supervisor is no longer needed (his or her presence or absence is undetermined) because the inmate exercises control over his or her behaviour as a result of the uncertainty of potentially being watched. Foucault describes this process of control as the **automatization of power** because

> the surveillance is permanent in its effect, even if it is discontinuous in its action; that the perfection of power should tend to render its actual exercise unnecessary; that this architectural apparatus should be a machine for creating and sustaining a power relation independent of the person who exercises it; in short, that the inmates should be caught up in a power situation of which they are themselves the bearers. (p. 201)

In these kinds of environments, individuals need to engage in complete self-censorship and regulation.

From this standpoint, the ideal form of discipline would "make it possible for a single gaze to see everything constantly" (p. 173). To some extent, this is similar to the metaphor of **Big Brother**, which became pervasive in the twentieth century to explain state control. This ideal form of discipline is a form of disindividualized power, where no one single observant is watching over the prisoner (Foucault, 1995). Rather, the architecture, location, and power relations create the sense of absolute control and constant surveillance.

While institutions and often buildings were set up to support various forms of surveillance, technology played only a minor role in enforcement. However, recent advances in digital technologies have given rise to new modes of data collection, storage, and retrieval. We discuss next the new modes of surveillance arising from the information revolution and contrast them with traditional forms.

Technology's Role in the New Surveillance

Information technology has made our lives more open to the public than ever before. New tools can easily collect, store, and retrieve personal information. Gary T. Marx (2007) identifies three distinct approaches to describing the changes in surveillance as a result of the information revolution and its associated technologies:

1. *Functional view*: Societies, in order to operate effectively, require some element of security and safety. To achieve these goals, personal

information needs to be collected and stored. From the functional view, the transformations in surveillance are only of degree, not of kind.

2. *Revolutionary view*: This is a rather pessimistic and deterministic view of the impact of information technology on surveillance. The revolutionary view argues that technologies have led to a radical transformation in the very nature of surveillance and that basic privacy rights have been jeopardized.

3. *Cultural view*: This view also sees information technology as radically changing society. But instead of advocating for a deterministic view, the cultural view argues that social and cultural factors moderate how information technology impacts surveillance. This view then sees counter-surveillance in combination with new privacy laws, norms, and values as leading toward a balance between disclosure and protection of personal data.

Although we can debate whether or not these technological changes should be labelled "fundamental" and "revolutionary" or "continuous" and "expected," there is no doubt that recent technological developments have created new tools and practices for surveillance. In addition, changes have occurred in our basic understanding of what surveillance is.

As Marx (2007) indicates, traditional definitions of surveillance may be restricted in their applicability to the new modes of surveillance that have emerged as a result of the information revolution. Traditional definitions tend to emphasize close observation of a "suspected person" (see, for example, the definition we discussed previously). By contrast, new surveillance is not limited to single individuals or suspects, and physical distance does not restrict one's ability to observe. In this sense, the definition needs to be broadened to accurately capture the nature of modern surveillance practices. Marx (2007) suggests that **new surveillance** can be defined as "the use of technical means to extract or create personal data" (p. 85). This definition emphasizes the use of technologies for the purpose of data collection, storage, and retrieval as a means of surveillance. To differentiate between traditional forms of surveillance and the new types of surveillance, Marx contrasts the key dimensions in Table 10.1.

While traditional forms of surveillance would often include informants or spies and wiretapping of telephone lines, new types of surveillance include hidden cameras in banks or stores, keystroke monitoring by employers, and audio scanners that pick up cellular phone frequencies. According to Marx's chart (Table 10.1), even something as seemingly routine as logging onto Facebook has elements of the new surveillance. For instance, personal information posted on Facebook may be collected by third-party companies and external websites (if the privacy settings allow for this type of access). This type of surveillance is largely invisible since Facebook members are often

Table 10.1 Dimensions of New and Old Surveillance

Dimension	A Traditional surveillance	B The new surveillance
Senses	Unaided senses	Extends senses
Visibility (of the actual collection, who does it, where, on whose behalf)	Visible	Less visible or invisible
Consent	Lower proportion involuntary	Higher proportion
Cost (per unit of data)	Expensive	Inexpensive
Location of data collectors/analyzers	On scene	Remote
Ethos	Harder (more coercive)	Softer (less coercive)
Integration	Data collection as separate activity	Data collections folded into routine activity
Data collector	Human, animal	Machine (wholly or partly automated)
Data resides	With the collector, stays local	With 3rd parties, often migrates
Timing	Single point or intermittent	Continuous (omnipresent)
Time period	Present	Past, present, future
Data availability	Frequent time lags	Real time availability
Object of data collection	Individual	Individual, categories of interest
Comprehensiveness	Single measure	Multiple measures
Context	Contextual	Acontextual
Depth	Less intensive	More intensive
Breadth	Less extensive	More extensive
Ratio of self to surveillant knowledge	Higher (what the surveillant knows, the subject probably knows as well)	Lower (surveillant knows things the subject doesn't)
Identifiability of object of surveillance	Emphasis on known individuals	Emphasis also on anonymous individuals, masses
Emphasis on	Individuals	Individual, network systems
Realism	Direct representation	Direct and simulation
Form	Single media (likely or narrative or numerical)	Multiple media (including video and/or audio)
Who collects data	Specialists	Specialists, role dispersal, self-monitoring
Data analysis	More difficult to organize store, retrieve, analyse	Easier to organize, store, retrieve, analyse
Data merging	Discrete non-combinable data (whether because of different format or location)	Easy to combine visual, auditory, text, numerical data
Data communication	More difficult to send, receive	Easier to send, receive

Source: Marx, G. T. (2007). What's new about the "new surveillance"? Classifying for change and continuity. In S.P. Hier & J. Greenberg (Eds.), *The surveillance studies reader*. Maidenhead: Open University Press, pp. 83–94. Table 6.1, p. 87.

not aware which companies are accessing their information and for what purposes. The companies that collect personal information from social network sites do so remotely and use machines (i.e., networked computers) to retrieve user data.

There is some controversy regarding the extent to which the use of personal data on social network sites represents a form of surveillance. We can examine this issue from the viewpoint of the three perspectives outlined on pages 198–9. From a functional view, we could argue that this form of surveillance is harmless since third-party companies are primarily interested in aggregate data and will use this information for the purpose of developing and marketing better products, which will benefit consumers in the long run.

These third party companies are not interested in individual behaviour, opinion, or attitude but, rather, in anonymous answers by masses of people and their trends and patterns.

Second, from a revolutionary view, we can take a more critical stance, arguing that most Facebook users grant third-party access to their personal information involuntarily by not being fully aware of their ability to change privacy settings or by not fully understanding the settings. The Privacy Commissioner of Canada criticized Facebook for having not only impossible to understand privacy rules but also rules that do not comply with Canadian law (Privacy Commissioner of Canada, 2009). Facebook's privacy rules have also come under scrutiny because they are longer than the US Constitution itself. Navigating this complexity of rules is not made much easier by Facebook's "Help Center," which is meant to assist users and has more than 45,000 words. As a result, users need to constantly balance the need to disclose personal information to connect with their family, friends, and co-workers while at the same time protecting themselves against privacy threats. Given the above, we can see how Facebook can be considered an example of the new surveillance, and with some reflection we can understand how pervasive surveillance truly is in our information society.

Finally, from a cultural view, we can argue that new laws are being put in place and new cultural practices are developing to guard users from privacy threats. For instance, in Canada the Privacy Commissioner demanded that Facebook change its privacy settings and make its options more transparent to allow users to make informed choices (Privacy Commissioner of Canada, 2009). The increased transparency should give users a better understanding of how their settings affect the kind of personal data that third-party companies and other users are able to access.

However, it is not always clear to what extent personal information garnered through surveillance is being used, especially in the area of government surveillance of its citizens. The evolution of government use of closed circuit televisions (CCTV) worldwide for the purposes of citizen surveillance is a case in point. While London, UK, is a leader in CCTV use, the city of "Chicago has deployed one of the most sophisticated networked systems, linking 1,500 cameras placed by police to thousands more installed by public and private operators in trains, buses, public housing projects, schools, businesses, and elsewhere" (Chesterman, 2011, p. 152). While there is no doubt as to the positive impact of such integration of technology, we must also wonder about its possible negative consequences. As noted by Alex W. Stedmon (2011), "as surveillance technologies become ubiquitous, the potential to monitor locations from distant control centres (possibly even from control centres abroad) means that gaps in spatial knowledge and lack of local knowledge could impact on many aspects of successful surveillance and public safety" (p. 533). As part of our discussion on new surveillance, we will now focus on how reality TV has drastically changed what surveillance means for society.

Reality TV

Reality TV has completely shifted the social meaning of surveillance. While traditional forms of surveillance stress the involuntary nature of the act of being watched by a single observer, in the context of reality TV, the Orwellian notion of Big Brother becomes reformulated (Andrejevic, 2004). In Orwell's novel *Nineteen Eighty-Four*, Big Brother is the metaphor for the state's complex surveillance apparatus. Technology plays a central role in reinforcing appropriate citizen behaviour by extending Big Brother's means of control. Television sets with embedded security cameras—referred to as telescreens—not only broadcast censored, state-approved content but are also interactive monitoring systems used to keep citizens under government control.

Orwell's Big Brother drastically contrasts with how surveillance occurs in the context of reality TV. In reality TV, millions of viewers watch the show *Big Brother* and its cast. Surveillance no longer is an infringement on privacy or a means of control but, rather, a mediated spectacle (Andrejevic, 2004). Participants welcome the chance to be part of the spectacle because it provides them with the unique opportunity to participate in the production of media.

In the past, a clear distinction was maintained between those who create media products and those who consume them (Andrejevic, 2004). In that model, audiences were passive recipients of information and had limited input in the development and production of content. Reality TV has shifted the locus of power, and audiences are now also active participants—they feel empowered through their direct involvement in the production process and the creation of content. Perhaps we can even argue that the participants of these shows *are* the content. Hence, reality TV reinvents the meaning given to surveillance because surveillance becomes yet another form of entertainment. Box 10.1 uses the reality series *Big Brother* to illustrate how the meaning of surveillance has changed in the context of reality TV.

Box 10.1 ❖ *Big Brother*: THE ACT OF BEING WATCHED

The reality series *Big Brother* debuted in 1999, first in the Netherlands, and since has become a success in at least 60 other countries. Even though the US version was not as successful as *Survivor*, it attracted a loyal audience. During the first US summer season, about 10 million viewers followed the show.

Andrejevic (2004) sees *Big Brother* as depicting a more authentic portrayal of real life than other reality series. The show revolves around the lives of 10 participants who live together in a house during the duration of the show. In the house, ironically, participants have no access to media (TV, radio, Internet); hence, participants have only minimal contact with the outside world. The idea is that participants will return to basics—engaging primarily in face-to-face contact with their fellow housemates. Every week, viewers (or, in the US version, housemates) vote to have one of the housemates evicted, causing that participant to lose the opportunity to win the big prize.

Big Brother does not share many elements with traditional forms of surveillance. Nonetheless, the show's setup does have a noteworthy resemblance to the architecture of the Panopticon. In *Big Brother*, cameras and microphones are located in every room behind one-way mirrors, even in bathrooms, providing a view into housemates' daily activities from different angles. Housemates are aware of the cameras, but they cannot see them. Like residents of the Panopticon, they know they are being monitored, but they cannot predict when. Moreover, the TV broadcast is heavily edited and cut, which means that any material could simply be disregarded. This puts the housemates in a position were they are constantly in what Goffman refers to as the **front stage**, that is, the public self (1959). And, in what Goffman referred to as the **back stage**, the housemates have very little private time. Moreover, the housemates do not control when their activities become public; as a result, they have the sense of always being on display. This resembles the power relations as described in Foucault's notion of the automatization of power and it also leads to a blurring of the self with capital means of production (Hearn, 2006).

This 24/7 surveillance is further enhanced by the additional and parallel coverage provided by the Internet. To obtain further details about the activities of housemates, online viewers can access a continuous 24-hour feed from multiple angles and rooms. The online version is an unedited live stream of the events as they unfold, in comparison to the heavily cut and edited TV broadcasts. The Internet stream is supplemented with parallel updates via email, text messaging, blogs, and discussion boards, leading toward a form of **hyper-surveillance** that to some extent reminds us of the state surveillance that was portrayed in the movie *The Lives of Others*. This compelling movie portrays how an agent of the secret police (STASI) is conducting surveillance on a well-known writer and his girlfriend. As the agent listens in to their most intimate conversations, he becomes increasingly absorbed and moved by their lives. The movie shows how surveillance can be all-encompassing and devastating for both those being observed and those conducting the surveillance. In reality TV, the observer is unknown and exerts control through the possibility of obtaining a glimpse into someone's "real" experience, that is, watching the ordinary.

Many people consider being a part of a reality TV show a move toward stardom and celebrity status. It is not just about portraying the ordinary life of audience members or about being oneself on TV as the label reality TV would suggest. By contrast, "the artificial reality portrayed in Big Brother, in which interactions are constructed, manipulated, monitored, and commodified, corresponds to the historical reality of artifice in a society permeated by commodified forms of entertainment, experience, and interaction" (Andrejevic, 2004, p. 149). *Big Brother* takes commodification to new levels, in that it sells everyday life to audiences, a process that has been referred to as the commodification of daily life. The inherent consequence of this is that "the promise of power sharing reveals itself as a ruse of economic rationalization and the promised form of participation comes to look at lot like work" (p. 7).

Digital Surveillance

The Internet is an information juggernaut, collecting a wide range of data on its users. The term *digital* surveillance refers not only to observation via the Internet but also to the collection of data via digital networks, tools, and devices. Some of the data collected is considered to be more personal and private than other types. Search engines, such as Google or Internet Explorer, track users' online search behaviour through search and toolbar logs—the logged data is referred to as the **search history**. The primary purpose of tracking this behaviour is to provide users with better search results as well as to display customized ads. Users do usually not consider these kinds of data to be personal and private even though many people do not approve of search engines logging their search history.

There are times, however, when users are asked to share more personal information. For example, when people subscribe to a new service on a website, they are often asked to provide their username and password; other websites collect more detailed information about users, including age, gender, location, etc. Even though these data are considered personal and private, most users will willingly disclose this information in order to gain access to the service (Greenberg, 2008) despite concerns about their privacy potentially being compromised. They view it as a minimal requirement in order to take advantage of the service.

A survey conducted by the **Pew Internet and American Life Project** shows that 43 per cent of Internet users know that search engines track their online search behaviour and as much as 57 per cent do not know this (Fallows, 2005). Most users, 55 per cent, disapprove of their online search behaviour being tracked while 37 per cent do not mind. Most users disapproved of websites displaying advertisements based on their search history and Web behaviour, which is interesting given that the majority of search engines rely heavily on this information. The following experience is not uncommon: a user sends an email to a friend about a roofing problem and shortly afterward ads for roofing companies are "coincidentally" displayed on that person's search page or email provider. Overall, the findings suggest that people are moderately concerned about the collection of personal information online and how this information is being used by companies. At the same time, many users do not fully understand how technology operates and the privacy implications of the tools they use, as demonstrated by the general lack of knowledge of how search engines operate.

How can we explain users' willingness to disclose personal information despite their privacy concerns? One perspective argues that online users are not fully aware of their vulnerability to privacy threats because they follow the "nothing to hide, nothing to fear" rule (Marx, G., 1996; Viseu, Clement, & Aspinall, 2004). Viseu, Clement, and Aspinall (2004) interviewed 10 Toronto Internet users about their privacy concerns when online. Those users who had not yet had negative consequences resulting from their online experiences were more likely to disclose truthful and accurate information. This suggests that for most users, online privacy is not a concern as they do not think privacy violation will affect them directly. Privacy becomes a concern only when it has been lost or breached.

A second reason to disclose information online is to have an online presence. For individuals to have an **online persona**, they must first engage in what Sundén (2003) has referred to as writing themselves into being. On social network sites, a number of elements constitute a user's online persona: the profile, status updates, pictures, and connections. As a result, the trend on the Internet is toward more disclosure of personal information; this has been defined in the literature as **information revelation** (Gross & Acquisti, 2005). Studies on information revelation have consistently shown that users of social network sites reveal considerable amounts of personal information on their profiles. In a study of undergraduate Facebook users, for example, 83 per cent of students revealed their hometown, 87 per cent revealed their high school, and 78 per cent disclosed their interests (Lampe, Ellison, & Steinfield, 2007). Box 10.2 discusses a study of the information revelation behaviours of Canadian undergraduate students as well as differences in gender.

The media has warned of the potential dangers of providing too much information online or the wrong kinds of information. Students have also expressed some concerns about the potential misuse of their personal data. Some of these concerns include a stranger finding out their address, their schedule, their sexual orientation, the name of their romantic partner, and their current political views (Gross & Acquisti, 2005).

Some theorists claim that interactive and digital media have led to the end of privacy as "the increased use of electronic communications has been matched by the development of ever more sophisticated tools of surveillance"

Box 10.2 ⚛ Information Revelation on Facebook

The need to belong and be an active participant in one's online community has led many young people to disclose personal information on social network sites. In a study of Facebook undergraduate users, Young and Quan-Haase (2009) found high levels of information revelation. Figure 10.1 shows the information indicated on respondents' profiles by gender. As much as 99 per cent reported using their actual name in their profile (first and last name). Nearly two-thirds of respondents indicated their sexual orientation, relationship status, and interests (such as favourite books, movies, and activities). Other personal information that most people included were their school name (97 per cent), email address (83 per cent), birth date (92 per cent), and the current city or town in which they live (80 per cent); almost all respondents reported posting an image of themselves (99 per cent) and photos of their friends (96 per cent). By contrast, few respondents reported disclosing their physical address (8 per cent), their cellphone number (10 per cent), or their IM screen name (16 per cent), thereby limiting the likelihood of individuals contacting or locating them outside of Facebook. For the most part, the data show that there was very little difference in terms of the types of information that female and male respondents include on their profiles.

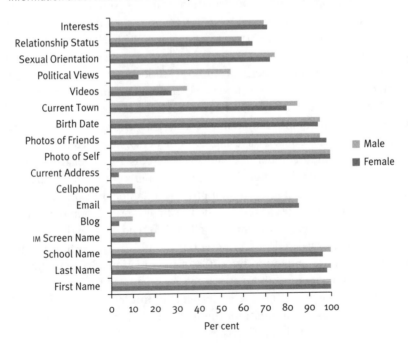

Figure 10.1 Information Provided on Profile by Gender
Source: Tufekci, Z. (2007). 'Can you see me now? Audience and disclosure regulation in online social network sites'. Bulletin of Science, Technology & Society, 28(1), 20–36, p. 31.

Students in the study reported that they are motivated to disclose information in order to engage with friends online. They see their profile as a signal that reveals aspects of their personality. These signals allow users to establish common ground and write their community into being (boyd, 2006). Young and Quan-Haase (2009) describe the importance of revealing information on Facebook for young people because it "seems to serve a social purpose increasing students' opportunities for social interaction and participation, as well as for the maintenance and formation of relationships" (p. 272).

(Chesterman, 2011, p. 3). There is no single definition of **privacy** because it is a fluid and far-reaching concept (Young, 2008). Westin (2003), however, has provided a useful definition that deals with individuals' personal information and defines information privacy as "the claim of an individual to determine what information about himself or herself should be known to others . . . when such information will be obtained and what uses will be made of it by others" (p. 431). This definition has pragmatic implications because it gives individuals control over their personal information and how it is used by other parties, including governments and corporations.

An important distinction made in the literature is between immediate and future privacy threats. Immediate threats are defined as those that result shortly after disclosing personal information on social network sites, including sexual predators and identity theft. Future threats are those that occur long after the information has been disclosed. A study of undergraduate students' privacy examined four types of future audiences: employer, romantic partner, government, and corporation (Tufekci, 2007). Table 10.2 shows that students thought the likelihood was higher that a future romantic partner would look at their profile than a future employer, the government, or a corporation. This suggests that students are most concerned about privacy

TABLE 10.2 **Gender Differences in Mean Levels of Perceived Likelihood of Being Found by Future Audiences**

	Total	Male	Female	t Test Statisitcs		
				t	df	p
Employer	2.90	2.90	3.10	−0.018	232	.986
Romantic partner	3.62	3.48	3.73	−1.404	232	.162
Government**	2.63	2.91	2.41	3.007	232	.003
Corporation*	2.55	2.76	2.39	2.264	232	.024

Note: Response options were: 1 = *never thought about it*, 2 = *not likely at all*, 3 = *a little likely*, 4 = *somewhat likely*, and 5= *very likely*. * *p* < .05, two-tailed. ***p* < .01, two-tailed. ****p* < .001, two-tailed.
Source: Tufekci, Z. (2007). Can you see me now?? Audience and disclosure regulation in online social network sites. *Bulletin of Science, Technology & Society*, 28(1), 20–36, Table 11.

breaches coming from their personal social network than from an institution or organization. Interestingly, students welcomed potential romantic partners seeing their profile because this could help them identify common interests and also provide an opportunity to get to know them better.

A recent study also distinguishes between **social privacy** and **institutional privacy** (Raynes-Goldie, 2010). Social privacy refers to the concern that known others—such as friends, acquaintances, and family members—will discover our personal information. By contrast, institutional privacy refers to the threat of information being mined and used by governments or corporations. In this study, users report being much more concerned about the loss of social privacy than institutional privacy. One of the respondents talks about "how frustrated he was with the 'context collision' created by Facebook's flattened Friend hierarchy, where by default, everyone is given access to the same personal information" (Raynes-Goldie, 2010). Google+ attempts to address this concern by allowing users to add people in distinct and non-overlapping circles. Through the use of circles, senders can target messages to specific groups of individuals, thereby overcoming privacy problems associated with context collision. Social and institutional privacy, then, have very different repercussions and social consequences.

There are several reasons why people are concerned about digital privacy threats.[3] First, digital content can easily be copied and forwarded online (boyd, 2007; Tufekci, 2007). Second, people worry about the persistence and searchability of the content, which, while making that content accessible to family, friends, and acquaintances, also allows access to strangers who have no connection with the author (Hogan & Quan-Haase, 2010; Young, 2008). Third, information taken out of context can be interpreted in different ways, potentially leading to misinterpretations. Fourth, often there is a clashing of different cultures, understandings, values, and norms when personal information is moved from one social setting to another (Barnes, 2006).

Taken together, the findings on users' behaviour on social network sites are puzzling. On the one hand, users report high levels of concern about the potential immediate or future misuse of their personal data. On the other hand, people continue to disclose large amounts of personal information, including pictures, information on friends, their whereabouts, etc. This contradictory behaviour has been termed the **privacy paradox** (ibid.).

Another paradox that emerges in social media is the nature of surveillance itself. Surveillance on social media is not about one person doing the observing and another person being observed, as in traditional surveillance; instead, surveillance on social media is about everyone being both object and subject at the same time. Tufekci (2007) has referred to this as peer monitoring because in most cases it is not strangers or spies who are watching and observing our behaviours; in social media, it is our family, friends, and acquaintances who are keeping a close eye on our every move. As a result, a social space has been created where we need to carefully craft our identity so that it fits with the norms, expectations, and values of our self-created audience (Tufekci, 2008). This process impacts our understanding of the creation and maintenance of social network audiences and the development and expression of identity.

Counter-surveillance as a Means of Personal Resistance

In an effort to counter the invasiveness of surveillance in our society, models of **counter-surveillance** have been initiated. In this section, we discuss two forms of counter-surveillance: **sousveillance** and privacy protection strategies.

Mann has described sousveillance as a form of counter-surveillance that empowers those subjected to institutional, state, and corporate surveillance practices (Mann, Nolan, & Wellman, 2003). In this approach, individuals use mobile technologies or wearable computers to record the experience of being watched. A subset of sousveillance practices are referred to as **inverse surveillance** and consist of recording, monitoring, analyzing, and questioning surveillance technologies and their proponents, and also recording how surveillance takes place by authority figures, such as policemen, guards, and border patrols. Around the globe, many groups have formed to participate in the sousveillance movement. For instance, those engaged in inverse surveillance tactics in New York City saw a 40 per cent increase in video cameras after the 11 September events (Mann et al., 2003). The pervasiveness of surveillance can be observed in many other cities, such as Chicago, London, and Washington.

Sousveillance entails **reflectionism,** a perspective that proposes using technology as a mirror to question and confront the ubiquity of surveillance in our modern society. The aim of reflectionism is to get people to think critically about how surveillance is occurring, thereby creating greater transparency. Reflectionism is a method for inquiry-in-performance with the aim of the following:

a. uncovering the Panopticon and undercutting its primacy and privilege;

b. relocating the relationship of the surveillance society within a more traditional notion of observability. (p. 333)

The process of reflectionism is closely linked with *detournement*, a concept introduced by Rogers (1994) to describe the tactic of using those very same tools that are employed to control us as a means to provoke the social controllers and make them aware of the imbalance of power. Reflectionism "extends the concept of detournement by using the tools against the organization, holding a mirror up to the establishment, and creating a symmetrical self-bureaucratization of the wearer" (Mann et al., 2003, p. 333). Anti-surveillance activists have formed groups and used various ways of reflectionism to make the public aware of the ubiquity of surveillance tools and their infringement on private and public life.

The second form of counter-surveillance, privacy strategies, occurs at a micro-level with people protecting themselves against potential threats to their privacy. As we mentioned earlier, the privacy paradox theory states that people tend to disclose large amounts of personal information online even though they express concerns about potential privacy risks. Recent evidence, however, suggests that people are not as naive and oblivious to threats as suggested in the privacy paradox theory.

A study of undergraduate students shows that they use a wide range of strategies to mitigate threats and are constantly managing their personal information online (Young & Quan-Haase, 2009). Table 10.3 shows the means and standard deviations for a series of items on Facebook users' privacy protection strategies. Respondents rated each item on a 5-point Likert scale ranging from 1 for "strongly disagree" to 5 for "strongly agree." The results show that the most frequently used privacy protection strategy was to exchange private email messages in Facebook to restrict access to the content. This strategy was followed by strategy 2: changing the default privacy settings on Facebook to restrict who can see what profile elements. Another frequently used strategy is to exclude personal information to restrict unknown others from gaining access. Interestingly, few participants have provided fake or inaccurate information on Facebook to restrict strangers from accessing their personal information. Students do not falsify information because their friends would question the validity of the information and wonder about its meaning.

TABLE 10.3 **Privacy Protection Strategies Used by Facebook Users**

Individual Items and Scale	Mean	Standard Deviation
1: I have sent private email messages within Facebook instead of posting messages to a friend's wall to restrict others from reading the message.	4.72	0.68
2: I have changed the default privacy settings activated by Facebook.	4.33	1.25
3: I have excluded personal information on Facebook to restrict people I don't know from gaining information about me.	4.08	1.17
4: I have untagged myself from images and/or videos posted by my contacts.	3.85	1.55
5: I have deleted messages posted to my Facebook wall to restrict others from viewing/reading the message.	3.64	1.55
6: Certain contacts on my Facebook site only have access to my limited profile.	3.47	1.70
7: I have blocked former contacts from contacting me and accessing my Facebook profile.	2.91	1.71
8: I have provided fake or inaccurate information on Facebook to restrict people I don't know from gaining information about me.	1.66	1.03

Note: Items were evaluated on a 5-point Likert scale ranging from 1 = "strongly disagree" to 5 = "strongly agree."
Source: Young, A.L., & Quan-Haase, A. (2012). Privacy protection strategies on Facebook. American Sociological Association, Denver, CO.

These privacy protection strategies are a way for users to balance the need to provide personal information on social network sites and be active participants in their online social networks—consisting of family, friends, and acquaintances—with the need to protect their personal data and restrict who has access to the content. Privacy protection strategies are becoming increasingly important in online environments, where private and public spaces easily blur, users can easily copy and transfer data, and personal information persistently leaves digital traces.

Conclusions

This chapter covered the complex topic of surveillance and compared traditional types with new modes of surveillance. Modern societies provide citizens of nation-states with many rights and opportunities. At the same time, these social systems are based on conventions of rationality, standardization, norms, and discipline that impose many restrictions on an individual's freedom. An additional complexity in these social systems is the role of technology in facilitating surveillance and imposing control mechanisms. Constant surveillance exposes vulnerabilities in individuals and may lead to a dystopic Big Brother world, where public and private life are closely monitored. We must recognize the pitfalls of technological determinism because surveillance technology alone does not determine how society defines and respects privacy.

Another central conclusion to draw from this chapter is that surveillance can take many forms. In the context of reality TV, surveillance is no longer perceived as a threat. The view of everyday life that reality TV provides is one that is not only welcome but that takes the shape of a spectacle, a form of entertainment. Surveillance—and the "ordinary" life that it provides a glimpse into—becomes a commodity that can be marketed as any other commercial good.

Finally, we conclude that the trend on the Internet is toward *more* disclosure of personal information. In part, this is a result of the proliferation and widespread use of social media, which further blurs the boundary of what is private space and what is public space. However, people are not completely oblivious to the risks associated with making so many aspects of their private lives public. Indeed, users of social media employ a series of strategies to protect themselves against privacy threats, including strategies of counter-surveillance, such as sousveillance and inverse surveillance. These strategies empower users and suggest that social norms, policies, and values shape how technologies affect privacy, creating a balance between disclosure and protection of personal data.

Questions for Critical Thought

1. Discuss the three factors—knowledge, growing impersonality, and enhanced control—that have played a role in the shift toward a Weberian society of rationalization and bureaucracy.

2. Define Foucault's concept of hierarchical observation and show how it plays out in the structure of the Panopticon.

3. Gary Marx puts forward three views as to how technology is changing society, namely functional, revolutionary, and cultural. Provide support for one of these views using Facebook as an example.

4. Describe the phenomenon of the privacy paradox in digital media as analysts have described it. How can we best explain users' paradoxical behaviour?

Suggestions for Further Reading

Andrejevic, M. (2007). *iSpy:* Surveillance *and power in the interactive era.* Lawrence: University Press of Kansas. The author discusses how interactive media, instead of liberalizing society (freeing it up from state control), provides the means for further social Taylorism through widespread surveillance and monitoring.

Foucault, M. (1977). *Discipline and punish: The birth of the prison.* New York: Pantheon. Through an analysis of prisons, this classic provides an in-depth historical study of the changes that have occurred in how society exerts control and enforces disciplinary action.

Chesterman, S. (2011). *One nation under* surveillance: *A new social contract to defend freedom without sacrificing liberty.* Oxford: Oxford University Press. This book provides an overview of the history of surveillance and provides many current examples.

Marx, G.T., & Steeves, V. (2010). From the beginning: Children as subjects and agents of surveillance. Surveillance & *Society, 7*(3/4), 192–230. This article critically questions the new forms of surveillance made available on the market, which allow parents and children to engage in mutual surveillance.

Tufekci, Z. (2008). Can you see me now? Audience and disclosure regulation in online social network sites. *Bulletin of Science, Technology and Society, 28*(20), 20–36. This study shows how surveillance is being reinvented in the context of social media.

Online Resources

Internet for Lawyers
www.netforlawyers.com/page/arizonas-jail-cam-makes-earthcamcoms-list-
25-most-interesting-sites
> Internet for Lawyers has an article about the use of "jail cams" and
> shows a video of people who have been brought into jail.

"Someone to Watch over Me"
http://youtu.be/pkjuy6WVYDO
> This video shows surveillance camera players, who aim to increase
> awareness of the ubiquity of street surveillance.

Surveillance Studies Centre, Queens University
www.sscqueens.org/
> This website contains useful information about projects and research in
> the area of surveillance.

CBC documentary *Peep Culture*
www.cbc.ca/documentaries/passionateeyeshowcase/2011/peepculture/
> This documentary provides an interesting overview of the recent
> developments in the phenomenon of peep culture.

11 Ethical Dimensions of Technology

Learning Objectives

- ⊛ To summarize and evaluate the three central themes of the book.
- ⊛ To examine the ethical and moral dimensions of our technological society.
- ⊛ To question the idea of the neutrality of technology.
- ⊛ To think critically about and present alternative perspectives to the metaphors of technology as destiny and technology as progress.

Introduction

In the previous chapters we have provided an overview of the many ways in which technology and society intersect. The primary goal of the final chapter is to summarize the three key themes that run through the book. First, the study of technology needs to be approached from a socio-technical viewpoint. Second, technology and innovation are closely interwoven with economics and hence have consequences for our understanding of inequality. Third, social change results inevitably from technological developments. The second goal of this concluding chapter is to discuss some of the ethical and moral dimensions that humans encounter as they engage with technology. This chapter explores the following themes: the neutrality of technology, technology as human destiny, and technology as progress. Through this discussion, the chapter emphasizes the unexpected consequences of technology with which society needs to come to terms. A good example is the political, social, economic, and human repercussions of nuclear technology—be it nuclear weapons or nuclear power plants. We use the distressing events at the Fukushima Daiichi Nuclear Power Plant, also known as Fukushima Daiichi, to illustrate the complex interplay of technological, social, economic, and human factors.

The Book's Three Central Themes

The book covers a wide range of technologies, their uses, and social implications. Nonetheless, we can identify three themes as most central in our analysis of society and technology. We discuss each theme next and draw some conclusions.

1. The Socio-technical Approach

Why is a socio-technical approach to the study of technology necessary? Early conceptualizations tended to focus on technology as material substance, disregarding the social nature of technological invention, implementation, and use (Feist et al., 2010). As society relies more heavily on technology, it is inevitable that we need a better understanding of the link between society and technology. The first attempt to examine this link was through the conceptual framework of technological determinism (Feenberg, 1999), which argues that technology is the single most important precursor of social change. This framework has been heavily criticized because it does not account for the complexity of the relationship between technology and society; it assumes, instead, that technology alone can exert change.

Recent conceptualizations have taken a socio-technical approach in which the social and technological are closely interwoven and mutually influence each other, which has been referred to as a mutual shaping process (Bijker, 2009; Bijker et al., 1999). In such an approach, technological, political, economic, cultural, and social factors are integrated to explain how social change occurs. That is, social change is no longer examined on a single dimension but, rather, is viewed as a coming together of different societal factors in a seamless Web.

Despite the strengths of a socio-technical approach for understanding how technology and society intersect, some limitations remain. First, the socio-technical approach does not explicitly delineate a select set of variables that need to be examined. The approach is rather vague in determining what factors are most central in a socio-technical analysis. Second, the approach does not state what mechanisms underlie the relationship between technology and society. It is unclear as yet exactly how these two forces come together and mutually shape each other. Finally, little detail is provided as to how a socio-technical approach should uncover the mechanisms underlying the mutual shaping process. While a wide range of methods have been proposed for the study of the socio-technical, vagueness remains as to what is the best way to approach such an investigation. These shortcomings suggest that much more work is needed to fill in the existing gaps in the socio-technical approach.

2. Technological Inequality

A central question investigated in this book is how technology and inequality are linked. Karl Marx (1970) was the first scholar to highlight how industrializing capitalism was closely connected with social transformations. Several developments—the Industrial Revolution, Taylorism, and Fordism—have all contributed to changes in labour processes, such as deskilling, mass production, and the moving assembly line. Schumpeter

(2004) later developed the Marxist theory of economic development by arguing that innovation was the driver of economic development because it created new markets and sources of revenue.

Based on Schumpeter's analysis, we can conclude that technological inequality occurs at three levels. First, the gap between those involved in innovation and those involved in the work force continues to grow. Bill Gates, Steve Jobs, and more recently Mark Zuckerberg are prime examples of this widening economic gap. Second, the difference in society between the haves and have-nots often plays out in terms of technological savviness. Those who lack technological skills are finding it increasingly difficult to join the workforce (McMullin, 2011), which creates divides along educational level, socio-economic status, and age, with older generations often not being able to find new employment as a result of their low levels of technological know-how. Third, a global divide exists between those nations who invent, produce, and distribute new technology and those who continue to fall behind.

A central conclusion of this book is that science and technology are key parts of the economy and that investing in R&D leads to social and economic advantages. Nonetheless, the example of Ireland—the Celtic Tiger, as described in Chapter 4—demonstrates that innovation is not easily achieved. Investment in R&D also creates vulnerabilities that are difficult to manage in an increasingly interconnected, global economy. Technological inequality is perhaps the single most important challenge of the twenty-first century. How do we distribute wealth and power? What role will technologies—such as the Internet and cellphones—play in the struggle for democracy and wealth? How can entire nations keep up with the information revolution? As technology continues to evolve and widely diffuse throughout society, it will become more essential to address these questions.

3. Social Change

While there is no single definition of **social change**, it is often described as major change of structured social action or social structure taking place in a society, community, or social group (Weinstein, 2010). Researchers have identified a number of dimensions of social change, including space (micro, meso, macro), time (short-, medium-, long-term), speed (slow, incremental, evolutionary versus fast, fundamental, revolutionary), direction (forward or backward), content (socio-cultural, psychological, sociological, organizational, anthropological, economic, and so forth), and impact (peaceful versus violent) (Servaes, 1999).

As Servaes notes, "The new traditions of discourse are characterized by a turn toward local communities as targets for research and debate, on the one hand, and the search for an understanding of the complex relationships between globalization and localization, on the other hand" (2010). In the context of technologically induced social change, analysts often preclude that these changes are negative. For instance, the introduction of machinery in the weaving industry resulted in job losses and was the start of a trend toward deskilling. Nonetheless, the term *social change* does not preclude positive change as in the context of development or **information and communication technology for development (ICT4D)**. In these instances, technology is used as an additional resource that facilitates the economic, cultural, and social development of social groups.

Social change does not always occur in predictable ways and simultaneously does not have the same implications for all members in society. As a result, the link between technology and social change is one that can only be examined in the context of specific social groups. This kind of analysis requires an in-depth understanding of multiple factors, including economic, social, cultural, and historical, that affect these groups. And even within bounded social groups, social change can occur differently for distinct social actors.

Ethical and Moral Dimensions of our Technological Society

Neutrality of Technology

A central concern in the debate about how technology intersects with society is the **neutrality of technology argument**. The proponents of this argument believe that technology is impartial because it lacks a set of moral values and direction. Within this model, technology is not an agent with moral choices but a passive object used to meet human needs and goals. This debate links directly to the definition provided in Chapter 1, where technology is equal to material substance. That is, technology is viewed as merely a means to an end. Consequently, technology itself is neither inherently good nor evil. However, what humans can exploit or accomplish with a given technology can fall into either category.

The value-neutral perspective suggests that since it is humans, and not technological artifacts, that possess a system of ethical, moral, and social values, technologies are only placed into a moral equation as a result of human use. Yet under what circumstances can technology accurately be described as being neutral? Swedish philosopher Sundström (1998) has described three instances in which technology could be deemed as being value-neutral:

1. *Multiple uses of tools*: The first argument is based on the notion that if a technology has multiple and ambiguous uses, value is assigned to that artifact only by humans through the possible uses of the device. Hence, the artifact does not limit its application or purpose to a singular function; rather the purpose for which it is used is completely open.

2. *Uncontextualized tool*: The second argument relies on the notion that value is assigned to an object only when a value-laden being, such as a human, opts to assign a certain set of values to the device through applied practice. Therefore, if a tool or technique is not used in practice by human beings, it will contain no value properties, due to "the neutrality *before* action and the neutrality of *inaction*" (p. 42).

3. *Tool as science*: The final argument is that since technology is the product of science, which is inherently neutral, technology should also be perceived as being value-neutral.

As technology becomes further entrenched into modernity, the notion of technological neutrality within cultural, social, ethical, and moral realms is increasingly rife with critical scrutiny. Much of the philosophical and scholarly debate over the neutrality of technology has centred on ideas of origin and design. Critics such as Feenberg (1991) have proposed that technology cannot be referred to as neutral because it is imbued with the values present in the particular culture or civilization from which it originated. Box 11.1 describes the development and worldwide deployment of the AK-47 rifle to illustrate how tool invention and use are closely interlinked with societal norms, values, and events.

Box 11.1 ☼ AK-47: QUESTIONING NEUTRALITY

The AK-47 rifle is the most widely used weapon worldwide. From Arctic battlefields to the jungles of Latin America and the deserts of the Middle East, the AK-47 is known for its rugged reliability, durability, and simplicity (Kahaner, 2007). The AK-47 consists of only eight pieces and can be quickly assembled in 60 seconds. As a result it can be easily mass produced, is cheap to manufacture, and is easy to operate. Additionally, the AK-47 is reliable in inhospitable conditions, is lightweight, and is effective as a killing machine. During the Cold War, these attributes made the AK-47 and its variants[1] the weapons of choice for the Soviet Union, the Eastern Bloc, and numerous armed revolutionaries in Latin America, Africa, and Southeast Asia.

Regarding the neutrality of technology, it can be said that the AK-47 is the number one actor in warfare today. From its inception, its aim was to kill. Despite this horrific mission, the social reality behind its existence is much more complex. Its designer, Mikhail Kalashnikov, was a Russian soldier injured by German gunfire during World War II. With the Soviet military eager to develop an assault rifle for fighting against the German army, competitions were developed for the design and manufacture of a reliable assault rifle capable of withstanding the freezing climate and muddy conditions of the Eastern front (Snezhurov, 2011). Following several modifications, the Soviet military successfully tested and adopted Kalashnikov's rifle in the late 1940s.

For Kalashnikov, the invention of the AK-47 was about the defense of an entire nation against the Germans, who were brutal and systematic in warfare. The AK-47 was designed to influence the war and help combat the German troops. While the AK-47 was not deployed during World War II, from the standpoint of the Russians and Kalashnikov himself the deployment of the AK-47 seemed like a necessary, and perhaps the only, means to combat the enemy.[2]

However, taken out of this social and historical context, the weapon becomes a source of sorrow, death, and never-ending wars. The ramifications of Kalashnikov's invention can be related to the traits that made it so popular. During the Cold War, many countries that did not receive military aid from the United States instead purchased weapons from the Soviet Union. Following the collapse of the Soviet Union, many of these weapons were sold on the black market to drug cartels, terrorist organizations, and other insurgent groups. Approximately 50 to 80 million AK-47s exist, many of which are counterfeited (Naim & Myers, 2005); as a result, the extensive supply has drastically reduced the cost for these weapons.[3]

Additionally, the ease with which the weapon can be operated has made it easy for child soldiers to be instructed about its use in conflict zones such as Sierra Leone. Figure 11.1 shows the extent to which this simple but enormously devastating weapon has expanded across all continents and regions of the world, leaving a profound social and political impact. The involvement of children in war makes the use of the AK-47 particularly contentious. Kalashnikov later opined that people should "blame the Nazi Germans for making me become a gun designer, I always wanted to construct agriculture machinery" (Roberts, 2007). The effect of the AK-47 on warfare has been such that Kalashnikov regrets having designed the weapon: "I would prefer to have invented a machine that people could use and that would help farmers with their work—for example a lawnmower" (Connolly, 2002).

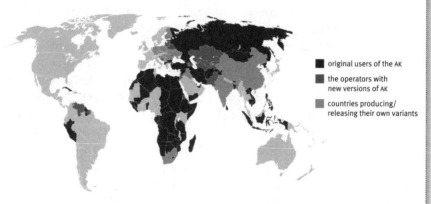

original users of the AK

the operators with new versions of AK

countries producing/ releasing their own variants

FIGURE 11.1 Geographical Spread of the Kalashnikov

Box 11.1 demonstrates the complexity of the neutrality debate. The AK-47 was originally built to defeat the German invasion of Russia. No one ever imagined that it would change the face of warfare for the worse. For Green, "[t]o argue that any technology is neutral is to ignore the social and cultural circumstances in which the technology was developed, and the policy and regulatory regimes under which that technology is deployed" (2002, p. 5). The social context in which a technology is developed does not have to match the context in which the technology is later utilized. Rather, technology is a reflection of society and its values, goals, and norms. Technology is not neutral because it embodies the power struggles between social groups and the societal goals that these groups pursue.

Pacey (1983) furthers this argument by noting that while a basic machine removed from its point of origin may appear at first glance to be culturally neutral, "the web of human activities surrounding the machine," including its symbolic status and uses, quickly negate its neutrality (p. 3). Viewing technology as an instrument open to human use and operating on a blank slate also overlooks the reality that the design and subsequent construction of any given technology can be "built in such a way that it produces a set of consequences logically and temporally *prior* to any of its professed uses" (Winner, 1999, p. 32).

Technology as Human Destiny

The metaphor of destiny in relation to technology is a powerful mode of approaching humanity's relationship to the world. During the Victorian era, the concept of destiny in relation to technology was heavily linked to the idea of progress. Inspired by Judeo-Christian historicism and Westernized interpretations of civilization, the Victorian conviction in linear progress was believed to be evidenced through technology. Following the carnage of the First World War, faith in a destiny of linear progress quickly evaporated. Factors such as the destructive consequences of the atomic age at Hiroshima and Nagasaki and the general increasing entrenchment of technology into the fabric of society greatly shaped the intellectual response to and surrounding discourse about humanity's technological destiny.

Jonas (2003) argued that if in the era of Napoleon destiny was achieved through politics, "we may well say today, 'Technology is destiny'" (2003, p. 14). Jonas divides technology into two distinct and separate spheres: **traditional technologies** and **modern technologies.** Traditional technologies have stationary and passive characteristics representing "a possession [and] a set of implements and skills" while modern technologies are an active "process [and] a dynamic thrust" (Jonas, 2003, p. 14). This is similar to Heidegger's (2010) comparison of traditional and modern technologies as seen through a hydroelectric plant on the River Rhine, which, unlike the windmills of yesteryear, transforms, controls, and manipulates natural matter in highly artificial ways. To confront the restless impulses of modern technology, Jonas argues for ethical responsibility due to "the central place it now occupies in human purpose" (1984, p. 9).

The relationship between technology, progress, and destiny also caught the interest of Canadian communication scholar George P. Grant. A fervent Canadian nationalist and philosopher, Grant's (1986) technological outlook was greatly shaped by the willingness of Lester B. Pearson and the Liberal Party to acquiesce to American requests to accept nuclear-capable warheads. Grant's broad definition of *technology* encompasses instruments and knowledge through which technology is an ontology, or mode of being connected to the nature of existence. He argues in favour of a view where technology is regarded as our "civilisational destiny" (p. 22), resulting in a view of destiny as an imposed mode of being, in which technology engulfs every aspect of society. The metaphor of destiny as portrayed by Grant suggests that human destiny is closely linked to technological change. To some extent it also suggests that technology is an inevitable and central part of humanity.

For Grant (2002), "faith in progress through technology" is such an ingrained and permanent component of North American life that "[t]he loss of this faith for the North American is equivalent to the loss of himself and the knowledge of how to live" (p. 400). For Grant (1986), this technological destiny directly shapes how we represent and make sense of the world around us and of ourselves. Grant (1986) is critical of the close interweaving of human subsistence and technology because as human reliance on technology increases, so consequently does the need to find technological solutions to correct and handle technological problems. As technology becomes weaved into the very fabric of human existence—material, cultural, economic, and social—humans will find it increasingly difficult to detach themselves from technology and comprehend its dangers. The more our destiny is one with technology, the more difficult it is for us to step back and take a critical standpoint.

Destiny is a central part of Heidegger's inquiry, which he describes as a "direction, which at best may be said to set a framework and provide a set of conditions" (Ihde, 2010, p. 38) for an eventual purpose or end. For Heidegger (2010), human destiny is not fully determined but is closely linked to human agency and choice. Modern technology endangers this freedom by concealing the full reality of its true nature. Deluded into believing they are masters of technology, humans may be unaware of this problem and misinterpret or ignore the dangers of and risks associated with technology. When humans fail to be concerned about technology and do not question its hidden intentions, Heidegger (2010) fears that humans will simply become an object of technology. This is an interesting point as it represents a reversal: technology is not an object that humans employ; instead, humans are objects, part of a technological system.

Heidegger (2010) offers a solution to the problem: to not outright reject technology but to detach ourselves from technology and extensively question its purpose and role in society—what he refers to as the essence of technology. This critical lens directly reflects his call toward viewing technology as an activity, rather than a series of artifacts designed for human means (Verbeek, 2005). Thus, Heidegger supports the argument that "modern technology is not destiny imposed upon humanity, but rather a manifestation of the effort by humanity to gain a measure of control over the forces of nature" (Zimmerman, 1990, p. 251).

Criticisms of Heidegger's outlook toward destiny have been principally drawn from scholars, such as Feenberg, who have rejected the fatalism of Heidegger's approach. The critics point toward two problems with Heidegger's view. First, technology is neither always nor only negative for humanity. Technology can improve the human condition in all realms. Second, the view of technology as faith is simplistic and denies human agency. Critiquing the works of Heidegger, Feenberg (1999) argues that his viewpoint produces conclusions in which "technology rigidifies into destiny . . . and the prospects for reform are narrowed to adjustments on the boundaries of the technical sphere" (p. 14). As a result, Heidegger's approach allows minimal space for the development of alternative technological practices, which perhaps take into account the moral and ethical dimensions of technological use. Additional criticisms of Heidegger's philosophy of technology have been connected to his abstract language and his nostalgia for a **pastoral era** (Verbeek, 2005). Heidegger's (2010) distinction between traditional and modern technologies is useful for understanding his nostalgia. He advocates in favour of traditional technologies, which in his view do not interfere with human nature. He rejects the use, and related social, political, moral, and ethical consequences, of modern technologies because he sees them as disturbing the natural equilibrium existent in nature.

The progressive betterment of humanity through technology has an appealing utopian sentiment, yet perhaps as philosophers such as Grant and Heidegger have argued such an ideal is also potentially fraught with naivety. If technology is a core function of humanity's destiny, to what extent is humankind able to extract itself from the force of technology? Additionally, to what degree should humans initiate a collective inquiry into the potential dangers, risks, and consequences (expected and unexpected) of a technology before its implementation?

For Cold War–era philosophers, questioning technology was necessary to prevent the likelihood of an autonomous or semi-autonomous technology exceeding the control of its human masters with damaging consequences, such as nuclear war (Ihde, 2010). In our contemporary Western societies, blind faith in the power and infallibility of technology is perhaps another area requiring critical inquiry and reflection. For example, on the surface modern technologies are seemingly able to master and control the earth's resources for human use, but in actuality these technologies are still highly vulnerable to nature's incalculable fluctuations and power. In Box 11.2, we discuss as an example the dangers and fears caused by the Fukushima Daiichi Nuclear Power Plant in Japan following the 2011 earthquake and tsunami that devastated the northeast coast of Japan.

BOX 11.2 ❖ FUKUSHIMA DAIICHI NUCLEAR POWER PLANT

Sadly, our ability to question our relationship to modern technologies and their potential risks seems to occur only following unforeseen disasters. In 2011, the Tōhoku earthquake and subsequent tsunami greatly affected the Japanese people. The most significant by-product of this natural disaster was the severe damage sustained by the Fukushima I Nuclear Power Plant. As the events unfolded, the media reported on what was known at the plant, the potential repercussions, and what political actions needed to be taken. The nuclear power plant disaster reveals the complexity of the technology. Even though nuclear particles are highly toxic, we cannot smell, feel, or see them. As a result, we have no means of detecting through our senses the dangers that nuclear particles present. For instance, one of the workers at the plant—often referred to as the Fukushima 50 Heroes[4]— died because he stepped into water contaminated with nuclear particles without wearing protective gear. The problem was that the workers were not even aware that the water surrounding the plant was contaminated.

In addition to the problems associated with nuclear technology, our control over nuclear plants is limited. As long as the plants are properly functioning, they present only a minimal risk to the population. However, the damage incurred through the Tōhoku earthquake shows that the technology can suddenly become very difficult to control. The news coverage demonstrated that little information was available as to exactly what was happening in the plants, with workers struggling to regain control of the situation.

The resulting equipment failures and emission of radioactive materials at the site brought about calls for a review of nuclear energy policies around the globe. Hibbs, a senior associate for the Carnegie Endowment's Nuclear Policy Program, proclaimed that the disaster "was a wake-up call for anyone who believed that, after 50 years of nuclear power in this world, we have figured it out and can go back to business as usual" (Eisler, Schmit, & Leinwand, 2011). Nevertheless, whether events like the Fukushima I crisis will truly be a "wake-up call" and encourage open and direct questioning of our technological destiny is yet to be seen.

As a result, certain countries, including Germany, have heeded the warning and decided to halt any new developments of nuclear power. Eventually, Germany's goal is to cease reliance on this potentially dangerous technology. Technological development, then, does not always move forward; rather, instances exist where a technology is abandoned for alternative and safer choices. In effect, human agency and societal attitudes can shape how a specific technology is used and whether or not it is discontinued.

Technology as Progress

Guiding our understanding of technology is the idea that it leads to progress. This idea emerged in the seventeenth and eighteenth centuries, particularly in Europe and North America, where technology was enabling radical economic, workplace, social, and cultural change. Technology assured "liberation and enrichment on the basis of the conquest of nature which was to be accomplished through the new natural sciences" (Borgmann, 1984, p. 216). Advances in the natural sciences allowed new technologies to be developed, and these created possibilities for how people could travel, work, and spend their leisure. An example is the creation of the railroad industry, which connected remote locations, thereby allowing for travel and hence the rapid transfer of ideas, goods, and people.

This notion of technology as progress is still deeply rooted in Western culture and continues to impact how we perceive, use, and evaluate technology. People view technology as a social and economic force and "[t]he dominance of particular technologies is often used to mark the 'progress' of modern western societies" (Hill, 1989, p. 33). These technologies become such markers of the times that we actually describe society in terms of the dominant technologies, for example, the industrial era, the atomic age, and the information society. In this view, then, technological inventions represent an improvement on previously existing technologies (Street, 1992), and progress can only be achieved if we continue to develop new technologies because without innovation society stagnates.

Baudrillard (2005) describes technologies as objects whose value is developed through our perception of their functional or symbolic worth. In his reflections on how technology and progress are linked, he contends that "technological society thrives on a tenacious myth, the myth of uninterrupted technical progress accompanied by a continuing **moral 'backwardness'** of man relative thereto" (p. 133). His notion of moral backwardness is essential to understand how individuals stand—vis-à-vis technology—as inferior entities who do not question the nature of their social system. Moreover, for him technological advancement does not represent progress but, rather, is characterized by a **model of regressiveness**. Technology is regressive because instead of society aiming toward moral progress, by questioning the present production system with its inequalities, power relations, and injustices, technological progress and failure is put at the forefront, perhaps as a means of distraction.

Baudrillard's (2005) theory uses the term **gizmo** to describe technologies that do not have a clear purpose in society. He describes how a gizmo is "always an indeterminate term with, in addition, the pejorative connotation of 'the thing without a name' . . . it suggests a vague and limitless functionality" (p. 123). Gizmos "serve to reinforce the belief that for every need there is a possible mechanical answer" and that "every practical (and even psychological) problem may be foreseen, forestalled, [and] resolved in advance by means of a technical object that is rational and adapted—perfectly adapted" (p. 125). Indeed, in a Westernized viewpoint, we can relate "the measure of civilization" to the abundance and complexity of available technologies (Street, 1992).

Additionally, this perspective stresses that technology is an enabler, giving people the means to fulfill their needs and wants, and it supports the idea of **technological utopianism**, in which society uses technology to create and maintain an idealized societal form (Segal, 1985). Supporters of technological utopianism have "made technological progress equivalent to progress itself rather than merely a means to progress, and they modeled their utopia after the machines and structures that made such technological progress probable" (pp. 74–75).

The view of technology as progress is rather myopic, as we discussed in Chapter 3, because it does not take into account the social, economic, and cultural consequences of technology. Simply put, progress is not a one-dimensional concept; technology and its use are embedded in the daily reality of people's everyday lives. Moreover, the metaphor of technology as progress leads us to a set of key ethical, moral, political, and social questions:

1. Is technological change always necessary?

2. Does technological advancement improve humanity?

3. Are there unaccounted-for consequences of technology that are not always apparent?

4. Does the technology-as-progress paradigm reflect capitalist notions of society?

These questions require careful consideration. They point toward implicit notions of how technology and society intersect. From this perspective, it is only through technological progress that society can solve its fundamental economic, social, health, and ethical problems. Technology, then, is no longer an option; rather, it is human fate because it is the only means through which humans can build a better world (Francis, 2009). As a result, people take technology for granted and no longer engage in questioning technology (Feenberg, 1999). Nevertheless, there are opposing opinions of this viewpoint because, for some, technological change "threatens established ways of life" and therefore they see such change as a regressive force (Street, 1992, pp. 20–21). We must continue to explore these questions as technology becomes ever more pervasive and ubiquitous in society.

Conclusions

In contemporary society, technologies and technological systems are embedded in the functions of our daily lives. Our interactions with these devices have become almost second nature—to the point that we think nothing of the interplay between ourselves and mechanisms as seemingly mundane as toasters, coffee makers, televisions, and computers. Since the beginning of the Industrial Revolution, people have viewed the escalation and sophistication of technologies as a mark of progress at both a technical and a social level. These devices have increasingly become more complex in their design and functionality.

Some consider technologies to be merely neutral: passive tools and techniques humans use to fulfill specific goals or needs. Others have strongly argued that the view of technologies as neutral entities is unrealistic and does not take into account the social, political, and cultural values of a particular technology's creators.

Naturally, as a society we will attest to the belief that most technologies have improved our lives: we can travel longer distances and access information resources via digital networks. However, critics may argue that while technologies have reduced our need to participate in physically exhaustive work, they have also changed the patterns of work in Western society. The economies of developed countries are typically driven by the tertiary service sector rather than by secondary manufacturing. As a result, workers must have much higher levels of education and technical sophistication in order to be qualified for and enter a professional career. Thus, while in one realm our lives are made easier by technology, in another, they are modified.

For theorists such as Martin Heidegger, a key to understanding the problems of technology was to be found in our ability to remove ourselves from technological objects and systems; to stand back and question technology's role in society. But is this even possible in modern society? Certainly, most people do not question the role and purpose of a technology unless that technology personally affects them in a profound manner. The raw emotion brought on by unforeseen disasters and tragedies is often a powerful catalyst that spurs a questioning of the role of technologies, such as guns or nuclear energy, in contemporary society.

Yet the ethical conundrum of contemporary technology is an important issue. As technology becomes further entrenched in the modus operandi of our twenty-first-century existence, humans must examine the short- and long-term effects of our relationship to technology. We do not call for a rejection or an abandonment of technology but for a measured evaluation of technology in order to maintain a healthy social, economic, and political relationship between ourselves and technology.

Questions for Critical Thought

1. Discuss the key arguments in favour of and against the neutrality of technology.

2. What are the problems underlying the metaphor of technology as progress?

3. Is it possible to critically examine technology if it is an intrinsic part of humanity's destiny? Provide three arguments to support your viewpoint.

Suggested Readings

Baudrillard, J. (2005) *The system of objects*. New York: Verso Press. An important philosophical work in understanding the value placed on objects within contemporary society.

Feenberg, A. (2002). *Transforming technology: A critical theory revisited*. New York: Oxford University Press. A classic work defining and crystallizing Feenberg's approach toward technology.

Grant, G.P. (1986). *Technology and justice*. Toronto: House of Anansi Press. This essay offers an important insight into George Grant's philosophical approach toward technology.

Heidegger, M. (2010). The question concerning technology. In Hanks, C. (Ed.) *Technology and values: Essential readings* (pp. 99–113). Malden, MA: Blackwell. Heidegger's essay is arguably the most significant and influential work concerning the philosophy of technology.

Online Resources

UNESCO's Ethics of Science and Technology Programme
www.unesco.org/new/en/social-and-human-sciences/themes/science-and-technology/news
Created with the establishment of the World Commission on the Ethics of Scientific Knowledge and Technology (COMEST), this site documents

UNESCO's attempt to create a dialogue for understanding science and technology within an ethical framework

3TU. Centre for Ethics and Technology
www.ethicsandtechnology.eu/
Based in the Netherlands, the 3TU. Centre for Ethics and Technology is a research-oriented collaboration by three Dutch universities dedicated to the study of ethics in science and technology. The site contains a publication database, areas of research, and a list of key members.

Glossary

access: When an individual has a means to connecting to the Internet, either through a computer or another digital tool. A number of different means of accessing the Internet exist, including Ethernet, dial-up, and wireless (WiFi).

actant: A non-human actor who engages in relationships with human and non-human actors.

actor: A person or entity bearing the capacity to (inter)act independently within society. The term *agent* is analogous.

actor network: In the context of ANT, this is the sum of actors and their complex Web of interconnections.

actor network theory (ANT): A sociological theory popularized in the 1980s by scholars Latour, Callon, and Law, which examines relationships between actors and envisions the world as a series of continuous and related webs.

actualistic studies: Archaeological studies using materials akin to those used in the past in order to provide links between behaviour and physical remains.

adhocracy: Where individuals or teams are assembled as they are needed to solve narrowly defined, short-term problems instead of having permanently assigned roles and functions based on organizational charts.

adoption (or **technological adoption**): The decision to use an innovation in order to facilitate the achievement of specific goals.

algorithm: A problem-solving method used in mathematics and computer science expressed in the form of a series of instructions.

alienation: A term used to describe the feelings among workers who, due to the standardization of their craft, begin to experience disconnection with and apathy toward their work duties.

angel investors: Individuals who provide capital investment for initiating technology businesses and in exchange become shareholders of the company or receive some form of repayment over time.

artifact: An archaeological term describing any object used, constructed, or modified by a human.

artificial intelligence (AI): A branch of computer science dedicated to designing machines capable of resembling or outperforming human intelligence.

asynchronous forms of communication: Types of communication that do not require each user to be present at the same time. Examples of asynchronous forms of communication include emails and text messages.

augmentation: The ability to supplement the physical human body by connecting it to digital components with computational capabilities.

automatic dialing: A form of telephone technology developed in the 1920s enabling users to automatically connect to other customers without having to go through a third party.

automatization of power: Foucault's term to describe how surveillance can be used as a means to enforce self-control.

autonomous: The capability of a technology to independently act within a selected environment. Within this view, technology is guided by its own internal logic, which directs and shapes social interactions and systems of thought.

autonomous Marxism: Also known as autonomism, this branch of Marxist theory emphasizes workers' ability to self-organize with the aim of creating changes in the workplace and throughout society at large.

autonomous technology theories: Philosophical or sociological approaches based on the belief that humans have little choice in deciding how a technology will evolve and diffuse in society. Technological determinism is an example of an autonomous technology theory.

awareness knowledge: Attained through awareness of the existence of a particular technology or innovation.

back stage: The elements of an individual's identity that are not revealed to the public but, rather, remain private.

basic research: Scientific inquiry aimed at the development of knowledge about the world.

Big Brother: A fictional character in George Orwell's novel *Nineteen Eighty-Four*. Big Brother, the dictator of the totalitarian state of Oceania, has come to represent control and surveillance for the purpose of maintaining power.

Big Brother: A reality-TV show that has been broadcasted in about 70 countries. Participants live in a house with minimal technology, and one of the housemates gets booted off the show every week.

black box of design: A term used to describe how the unfolding of technological invention and development is difficult to observe.

blogs: Originally referred to as Web logs, a blog is a type of website or a portion of a website that typically contains personal or informal views, media, and commentaries, which are displayed in reverse-chronological order.

breakup 2.0: The dissolution of a romantic relationship and the role played by Web 2.0 technologies, such as Facebook, Twitter, and Flickr, during and after the dissolution.

bridging connections: External relationships or communications for the purpose of bridging organizational boundaries to assist in the creation of boundary-spanning structures.

bureaucratization: A social system that allows organizations—both public and private—to achieve their goals through the implementation of rules, norms, and values.

capitalism: An economic system where the means of production are privately owned.

cellular network: A radio network consisting of cells, which are served by individual transmitter towers, or cell sites.

Celtic Tiger: A nickname used to describe Ireland as it showed rapid economic development linked to information technology.

change agents: Individuals who make potential users aware of an innovation and assist in providing information designed to help make informed decisions about the adoption of an innovation.

Chappe communication system: A form of pre-electric visual communication developed by Claude Chappe that relied on transmitting messages via towers. The system was based on a semaphore relay station, which allows conveying messages through visual information.

chatterbots: Computer programs designed to simulate a conversation with human participants through textual or aural means. See also *intelligent agents*.

C-Leg: A prosthetic knee-joint system, using hydraulic components, that is able to adjust itself to the changing speed and walking conditions of its user.

closure: Occurs when a social group has finalized experimenting with a new tool, as no new meanings or uses are ascribed to this artifact.

closure by redefinition of the problem: When the meaning of an artifact is established by rethinking the original problem instead of making changes to the artifact itself; for instance, in terms of its function or appearance.

Community Access Program (CAP): A Canadian government initiative, administered through Industry Canada, aimed at providing Internet access and skills for rural Canadians.

community-liberated view: Where community life is not lost but has changed with people socializing outside their local neighbourhoods and their immediate family ties.

community-lost view: Sees industrialization as responsible for a decline in the prosperity of community.

"community question": Addresses issues around how community has changed over time.

community-saved view: Where friendship and family ties continue to exist and form close-knit clusters similar to those found in the pre-industrial era.

compatibility: The extent to which an innovation fits with a social group's existing norms, values, and attitudes.

complexity: A characteristic in the model of adoption that describes the level of proficiency people need in order to understand how a particular technology works.

computer-mediated communication (CMC): Forms of communication that require the use of computer technologies, such as the Internet.

confirmation stage: The period in which potential adopters continue to seek out information about an innovation in order to ascertain whether or not they have made the right decision.

content of thought: The changing manner in which humans were portrayed as a result of the shift from an oral to a literate culture.

cool media: A term used by McLuhan to describe forms of media that require greater effort on the part of the viewer to understand the content and determine meaning.

core team: A cohesive group of programmers who work closely together to develop a software product.

counter-surveillance: An umbrella term used to denote any form of resistance against surveillance, and includes ways of evading surveillance by the state, increasing awareness of state surveillance, and acts against the mainstream forms of surveillance.

creative destruction: A term Schumpeter uses to summarize the social, economic, and cultural transformations that occur as a result of innovations.

creative processes: The term Schumpeter uses to describe the design, development, and implementation of new technologies.

creeping: The act of looking at others' profiles, wall posts, and pictures on social network sites, such as Facebook.

critical problems: In the context of system theory, these are complex problems that require socio-technical solutions.

cybertariats: Presented by Huws as individuals who are employed in basic data entry jobs for minimum wage.

cyborg: A being or entity containing both artificial and biological components that are seamlessly connected.

cyclical fluctuations: Schumpeter's term to describe changes in the economy resulting from innovations appearing simultaneously in clusters, instead of being spread over a longer period of time.

decentralized decision making: A process that allows workers rather than management to be at the forefront of decision making and information exchange.

decision stage: A period featuring the activities that lead toward the adoption or rejection of an innovation.

dehumanization: A concept frequently featured in the work of Ellul that refers to the manner in which technology has engulfed every level of society and human existence at the latter's expense.

democratic divide: Describes the differences in political engagement of those who have access to the Internet and those who do not.

deskilling: The elimination, reduction, or downgrading of skilled labour due to the introduction of technologies within the workplace.

developed country or nation: A term used to describe a nation with a high level of social, technical, industrial, material, and economic development. A list of developed countries can be found in the *International Monetary Fund's World Economic Outlook Report*, April 2010.

developing country or nation: Nations with a low level of material well-being and lacking a high level of social, technical, industrial, material, and economic development. The term is not to be confused with *Third World countries*, which historically has a very different meaning.

deviant behaviour: Behaviour deemed unacceptable when judged against the established norms of a society.

diffusion of innovations: The study of the adoption and spread of technological innovations in society, or segments of society, over time.

digital divide: Describes discrepancies between social groups in access to, use of, and empowerment by networked computers and other digital tools, such as cellphones, PDAs, and MP3s. The term also can encompass differences in skill level and knowledge about digital artifacts.

digital immigrants: The generation that did not grow up with the Internet and only started using digital technologies in their adult years.

digital natives: A term used to describe people born during or after the development of digital technologies who have correspondingly grown up with a knowledge and familiarity of digital technologies.

digital public sphere: Public spaces formed online to help citizens organize and mobilize.

digital revolutionaries: Protest leaders who use the Internet as a tool to organize protests and inform and mobilize citizens.

digs: Archaeological sites that have been excavated.

discipline: A central concept in Foucault's analysis of power. He describes discipline as the technique used against individuals to exercise power as well as the actual instruments used to enforce it.

discontinuance: The rejection of a technology or innovation after initial adoption and previous use.

domestication of the Web: The use of personal computer and Web-based applications for everyday personal, social, or leisure purposes and interactions.

dystopian perspective of technology: The belief that technology has a destructive or negative influence on society. This can include the re-organization or changing of social or labour practices due the imposition of technology.

dystopian perspective of the Internet: Predicts negative social consequences as a result of Internet use, in particular the weakening of social relations and communities.

ease of invention: Describes the extent to which an innovation can be easily discovered on the basis of existing knowledge. Some inventions are difficult to discover because they require not only complex knowledge in an area but also the combination of knowledge from distinct areas.

economic opportunity divide: Reflects beliefs and attitudes that individuals have about the advantages provided by access, such as finding a job, obtaining health information, and being able to take an online course.

end-users: Individuals who access websites or social media applications. Depending on their level of engagement, individuals may also supply, modify, or critique content within a Web 2.0 environment.

entrepreneurial spirit: Describes the attitude of many IT workers, who are eager to start new companies—despite the risks involved in the industry—because of the potential to obtain high returns on investments (ROI).

epigraphy: The study of the history and social circumstances of written information that has been preserved on hard materials.

equipotentiality: A core component of produsage projects that highlights the ability of all participating members to contribute to the end result.

E-Rate: A program officially known as the Schools and Libraries Program of the Universal Service Fund, which was established under the Federal Communications Commission (FCC). The aim is to make affordable Internet access possible for schools and libraries in the United States.

evolutionary model of technological development: An idea proposed by George Basalla, which suggests that new technologies arise from earlier sources, rather than from mere ingenuity.

examination: The process whereby information on individuals is recorded in documents, which serve the purpose of control as well as the establishment of norms.

excavation: The systematic process involving the exposure and recording of remains at archaeological sites.

Facebook: A popular social networking site, where people can create a profile and link to other users.

fixed lines: Also known as landlines; refers to telephone devices or systems that are operated in fixed locations.

Fordism: Refers to a system welding the principles of scientific management to standardize work processes.

Fordist culture: Describes the postwar mass culture of the 1950s and 1960s in which items were mass produced in a homogenized manner to appeal to the lowest common denominator.

friending: The act of requesting someone's affiliation on a social networking site.

friendship: Voluntary and informal relationships established between two or more parties who have interacted socially with one another. Whereas traditional friendships occur between friends who have shared localized experiences together, contemporary friendships may develop across a virtual sphere between individuals who interact via the Internet.

Friendster: A social networking site similar to Facebook that allows members to share messages and media.

front stage: Those elements of an individual's identity that are revealed in public.

gatekeepers: Individuals who bring information about a new innovation or technology into a social group.

Gemeinschaft: Generally translated as "community" and refers to a cohesive social entity that is united by pre-existing social bonds.

generalized reciprocity: Expressed in the notion of doing favours for and providing help to others without expecting them to return the favour.

Gesellschaft: Translated as "society" or "association" and describes the coexistence of individuals who are self-serving units and come together because of an over-arching goal.

gizmo: Baudrillard's term to characterize a technology that lacks a meaningful purpose within society.

global digital divide: Differences in access to the Internet and other digital tools among nations and regions.

globalization: The move toward the interconnectedness of human affairs—economic, cultural, social, and political—transcending national boundaries, governments, and laws.

global village: A term popularized by McLuhan to describe the possibility that electronic forms of media and communication have compressed spatial distances by enabling people to remain connected to activities and individuals throughout the world.

GNU General Public License: Stallman wrote this software license for the GNU project to regulate the use of code. It is the first copyleft license for general use of content on the Web. Copyleft is a type of licensing that is based on copyright law to allow for the use, the reuse, and the distribution of versions of an artifact. The same rights that apply to the original work also apply to all modified versions of that work.

goal: A term used by Heidegger to represent the desired end-product to which an act is originally directed.

Great Man Theory: A popular nineteenth-century theoretical idea popularized by Carlyle that supposes history can be largely explained and understood through the impact of notable individual leaders and heroes, who were often predominately male.

gross domestic expenditure on R&D (GERD): A measure the OECD uses that takes into consideration the total expenditure (current and capital) on R&D accrued by a nation over a one-year period, including expenses by companies, research institutes, universities, and government laboratories.

haves and have-nots: In connection with the digital divide, the "haves" relate to those who have the ability and the skills to access information and communication technologies, while the "have-nots" relate to those who, while lacking the ability to access these technologies, may demonstrate an interest in obtaining the skills related to information literacy or technical competence.

hegemonic elite: A dominant political, social, economic or cultural group that exerts power and influence over the remaining members of a society.

hierarchical observation: Describes how people's behaviour can be controlled through simple observation by those in positions of power.

hierarchical organizations: Entities or organizations that operate using a systematic power arrangement that positions one single individual or group at the top with the remaining members or participants in various levels of subordination.

hieroglyphs: Simple writing forms where graphical figures are used to depict words or objects.

high-definition media: A term McLuhan used to describe media that provide their users with well-defined and detailed data.

holoptism: An open approach to problem solving in which all participating members have equal insight into the problem's components.

hominin material culture: artifacts used by the ancestors of modern humans.

hot media: A term used by McLuhan to describe forms of media that require little effort on the part of viewers to process and understand the content.

how-to knowledge: An understanding of how an innovation is properly used and employed.

human agency: The belief that humans have the ability to make, choose, shape, and act upon decisions that have a genuine impact on the world.

human controlled: The ability of humans to control, design, and shape the actions of a technology or technological system.

hunter-gatherer societies: Communities or societies in which the primary method of subsistence involves having its members obtain food by way of hunting animals or gathering edible plants without participating in the domestication of either.

hyper-post-Fordist culture: The ability of people to customize and individualize their lives through consumer culture.

hyper-surveillance: Extreme forms of surveillance where many different kinds of personal data on an individual are aggregated.

idioms of practice: The social norms that evolve over time around media use.

imitation (or cyclical fluctuations): In Schumpeter's model of economic development, innovation does not occur evenly spread over time but, rather, in spurts or episodes.

immaterial labour: A form of labour that creates products or services that are not tangible and readily observable, such as knowledge or information.

impermanent record: A form of ancient writing that recorded daily life and because of its everyday nature was not meant to be preserved.

implementation stage: The period in which an individual begins to use an innovation.

income quartile: A statistical method used to categorize the population into groups based on average household income from poorest to richest. The division is based on the distribution of incomes, with each quartile representing 25 per cent of the overall population.

Industrial Revolution: A historical period beginning in eighteenth-century Western Europe spearheaded by changes in manufacturing, transportation, and technology, which markedly affected the production and distribution of goods as well as socioeconomic conditions.

inequality: Differences in society in terms of access to social goods, such as the labour market, income, education, and the health-care systems, as well as means of political participation. The discrepancies exist along socially defined grouping, including gender, age, social class, and ethnicity.

information: In Rogers' diffusion of innovations model, it is knowledge, fact, or advice communicated to an individual about an innovation.

information age: Also known as the computer age; relates to an era in which individuals are able to access and manipulate information with a high degree of sophistication and speed through the advent of information and communication technologies, such as computers, the Internet, and cellphones.

informational production or **informational**: A term used by Castells and Halls to describe an economy where economic development depends directly on the flow and processing of information and, ultimately, on the creation of new knowledge.

information and communication technologies for development (ICT4D): Describes the use of technologies that facilitate communication, the storage and reuse of information, and data analysis for the purpose of improving the economic and social conditions in poor countries.

information commons: A system, community, or institution designed to develop, store, share, and conserve knowledge and information. Physical libraries represent a traditional model of an information commons. In contemporary use, the term can also refer to digital services and/or virtual spaces that promote and adhere to these ideals, such as Wikipedia.

information literacy: The ability to successfully utilize, navigate, and operate information resources and information and communication technologies.

information revelation: The disclosure of personal information on the Internet, for instance on social network sites.

information revolution: The changes that have occurred in the types of technologies needed and their related services and products as a result of the shift toward the information society.

information society: A society in which the utilization, dissemination, diffusion, and development of information is an integral component in social, economic, and cultural life.

information systems development: Customized software development, where software is designed to meet the needs and requirements of a particular client or user group.

information technology (IT): The development and application of computer-based technologies designed for the principal purpose of creating, exchanging, and storing electronic information.

innerworldly asceticism: Refers to the Puritan ethic of self-discipline and self-control and its associated behaviours, norms, and values.

innovation: An idea, practice, or object that members of a social group perceive as being new.

innovator-entrepreneur: In accordance with Schumpeter's model of economic development, the innovator-entrepreneur is the most important driver of economic growth. This is an individual who develops new ideas, practices, or objects despite the risks and uncertainties associated with these innovations.

institutional privacy: Protection of personal information from companies and institutions (e.g., access to personal information by Facebook and its partners).

instrumentalism: A theory that analyzes technology as neutral tools or instruments whose purpose is to fulfill the specific tasks of their user.

integrated circuit: Usually referred to as a *microchip* or just simply *chip*, it is an electronic circuit built on top of semiconductor material. Each integrated circuit contains large numbers of minute transistors.

intellectualization: A term introduced by Weber to describe a shift in how decisions are made; instead of relying on superstitious or mystical beliefs, decisions are based on knowledge and rational choice.

intelligent agents: Entities capable of accruing knowledge and displaying intelligence to achieve goals related to their environment.

interactivity: The ability of users to provide content in response to a source or communication partner.

interconnectedness: The extent to which social groups are geographically or socially connected to one another, allowing for the flow of goods and the exchange of ideas, information, knowledge, and innovations. A high degree of interconnectedness allows for the easy transfer of technologies and technological know-how.

interconnectors: A role serviced by the early majority, who act as a bridge between early and late adopters and provide relevant information on new innovations to the latter.

interdependence: Describes the high level of connectivity currently existent in the world, which makes events that happen in one part of the world have a short- and long-term trigger-down effect on all other parts.

International Telecommunications Union (ITU): An international agency linked to the United Nations (UN) and concerned with questions about information and communication technology.

Internet censorship: Measures used by various countries to control the information that flows both into and out of a country, radically undermining citizens' ability to access and distribute information.

Internet dictator's dilemma: The dilemma confronted by oppressive regimes who on the one hand want to be active participants in the information society and on the other hand are concerned about the political repercussions of the open exchange of information and communication.

Internet kill switch: Refers to the act of shutting down the Internet to control access to information and communication primarily by governments.

interpretive flexibility: This term refers to how the meaning of artifacts is always created in a socio-cultural context.

invention: Considered a creative process, in which a person discovers a new behaviour, way of understanding, process, or object. Contrary to *innovation*, where a completely new idea is developed, in the case of invention the process consists of combining already existing elements in unexpected ways.

inverse surveillance: Practices that can be subsumed under sousveillance, but that emphasize surveillance as inquiry into social norms and expectation; involves the recording, monitoring, analyzing, and questioning of surveillance technologies and their proponents, and also the recording of how surveillance takes place by authority figures, such as police officers, guards, and border patrols.

invisible audience: Audiences that are not known to the creator of content when content is posted on the Internet.

iron cage: A concept introduced by Weber to describe how society through the introduction of rules of rationalization and bureaucratization has limited the ability of individuals to make their own choices and exert free will.

killer app (or killer application): Describes a computer program that people consider to be essential and valuable, and results in their purchasing the technology on which the program runs.

knowledge stage: In Rogers' model of the diffusion of innovations, this is the period in which an individual learns about an innovation for the first time.

latent ties: Latent ties consist of those social relations—including friends, acquaintances, and family—with whom we are not currently communicating or interacting but with whom we could reinitiate contact at any given point in time.

Layton's model of technology: An idea of technological development in which ideas are translated into designs. In the model, technology is embodied in ideas, techniques, and design.

least developed country (LDC) nations: A name given to countries that embody low socio-economic development due to a lack of human resources and economic resources, and to instable political structures.

literate societies: Societies that use writing as a means to preserve information as compared to oral societies, which rely on oral communication.

LiveJournal: A popular free social media site that allows users to keep a blog.

local virtualities: Geographically bounded places where individuals use computer-mediated communication to facilitate the exchange of information and the formation of social relationships.

low-definition media: A term McLuhan used to describe media that provide users with less information, requiring them to fill in the blanks.

Luddites: A social movement that emerged in nineteenth-century England, formed by textile artisans who protested against the increased mechanization of labour through sabotaging industrial machines and sites.

machine breaking: The process of sabotaging or destroying industrial machinery in response to inferior working conditions and the threat posed by mechanization to workers' crafts or livelihood.

mainstreaming of the Web: The notion that the Internet has become an integral component of society, including the development of social phenomena and relationships.

markets: An economic infrastructure or system that allows and encourages the practice of selling or exchanging goods and services for payment.

Marxist tradition: A critical theoretical approach to society based on the writings and ideas of Karl Marx.

mass consumption: A socio-economic system that nurtures a yearning to purchase increasing amounts of mass-produced goods under the auspices of social improvement and personal gratification.

mass production: The production of large quantities of standardized materials, made possible by streamlining the manufacturing process and adopting technology such as the assembly line.

material substance: Viewing technology as being its physical or material properties separate from its interactions with society.

mechanism: Designated by Heidegger to represent the means through which to achieve a goal.

media bias: Innis's term to describe the effect of media on how societies operate. He distinguished between a space and a time bias.

media ideologies: A term Gershon used to describe the beliefs users form about media and how these affect their use.

the medium is the message: An aphorism coined by Marshall McLuhan that proposes that the type of medium through which an individual encounters content may affect the manner in which an individual understands this content. McLuhan underlined the need to examine the impact of media on people instead of focusing only on the influence of content.

megacities: Metropolitan areas containing more than 10 million inhabitants, such as New York City.

model of regressiveness: Baudrillard's term for the manner in which society, distracted by ideas of technological progress, morally degenerates due to a lack of critical discussion about social inequalities and injustices brought about by the system of production.

modern technologies: Technologies that affect nature and human kind in radical ways; they have an active process and a dynamic impetus.

moral backwardness: Baudrillard's term to indicate the stagnation of morality brought about by ideas of technological progress and the trappings of materialism.

moving assembly line: A sequential manufacturing process that enables parts to be assembled within a specific order to create a product. This process was made famous by Henry Ford's production line, which allowed for the mass assemblage of automobiles.

multiple factor principle: A move away from seeing the inventor as the single cause for technological development. Instead, multiple factors are considered important in the process of innovation.

mutual shaping: A term used in STS to describe the close interrelation that exists between social and technological factors.

MySpace: A pioneering social networking site that prior to the advent of Facebook was the most popular social networking site in the United States.

National Telecommunications and Information Administration (NTIA): An agency of the US Department of Commerce whose information-based functions and policies relate to issues of telecommunications access and development.

nation-state: Self-identified and sovereign entities that cover a territorial unit, integrating individuals who live in a specific geographical area and who share an identity.

Ned Ludd: A weaver and mythical leader of the Luddites, who, according to folklore, destroyed his knitting frame in response to ill-treatment from his master. Ludd's name was adopted by workers who resented the imposition of technology in their places of work.

Neolithic period: An era of technological development beginning around 9500 BCE characterized by the transition from hunter-gatherer societies, which encompasses the advent of farming and settlement practices and ends with the widespread employment of metal tools.

Neolithic Revolution: The first agricultural revolution, which precipitated the shift from hunter-gatherer societies toward settlement and agricultural development.

Neo-Luddism: Individuals or groups who either openly reject or critique the role of technology in society.

netiquette: The social norms around communication and behaviour that have developed on the Internet. These norms are also linked to how language on the Net is unique, as exemplified by the use of acronyms such as LOL.

network: A term used by Latour to denote relationships or interactions between people, objects, and organizations whose shared purpose transforms these connections into an extensive collection of resources and participants that enables them to function as a single unit.

networked individualism: Sees society as moving away from a model where people are embedded in groups toward more loosely connected social networks.

networked organization: New forms of management that do not rely on a hierarchical structure of organization to achieve organizational goals; instead, employees are loosely associated.

neutral point of view (NPOV): A guiding principle of Wikipedia that stresses the expectation that participants will input information that is reliable and objective.

neutrality of technology argument: Technology is considered impartial because it lacks a set of moral values and direction.

newly industrializing nations: Nations whose economic, technical, and social development is outpacing developing nations, particularly through an emerging industrial sector. China and India represent newly industrializing nations.

new surveillance: The application of technical means to collect and record personal data.

nomadic tribes: Communities of people who prefer to roam from one area to another rather than settling in a single specific location.

nondiffusion: When an innovation or technology fails to be adopted by a social group.

normalizing judgment: A term Foucault used to describe how individuals are evaluated against set societal standards to determine if they can be considered normal (or are categorized as abnormal).

observability: The visibility of an innovation and its effects on members of a social group. By observing the innovation, members are more likely to discuss and gain information about it.

One Laptop per Child (OLPC): A US non-profit organization whose guiding principle is to create educational opportunities for the world's poorest children through information technologies.

online communities: Virtual communities whose members interact primarily online via synchronous or asynchronous forms of communication.

online persona: The virtual self projected by individuals through their online activities, such as their avatar, post history, status updates, and so forth.

open area excavation: A method of archaeological excavation involving the opening up of large areas of the site without baulks.

opinion leaders: Individuals within the early adopter category whose influence and opinions help guide the decision-making process of others in the social group.

optical telegraph: A method of transmitting information through visual signals or cues. An example of this would be smoke signals.

oral societies: Non-literate societies that preserve culture, knowledge, and norms via oral means.

Organisation for Economic Co-operation and Development (OECD): An international organization, consisting of 34 country members, that uses evidence-based analysis to promote policies on economic development and trade, with the aim of improving the economic and social well-being of people around the world.

organizational chart: A diagram featuring an organization's structure, rankings, levels, and components.

organization of thought: Secondary element determined by Havelock to be a fundamental response in the shift from an oral to literate society, which relates to how main and subordinate ideas are connected.

packaged software development: Software that is produced in large quantities and can be obtained off the shelf.

paleoanthropological record: A term denoting all evidence relating to the study of ancient humans, known as paleoanthropology; such evidence is sought out in an attempt to analyze the past.

Panopticon: A building that functions as a prison, providing complete control over the inmates. It also denotes an entire social system based on techniques of control through surveillance.

para-social grieving: The act of grieving for an individual with whom one has established a connection through media.

para-social interaction: A form of one-sided social interaction mediated via technology in which one individual has developed a personal bond with the other. This type of relationship is evident in the bond between celebrities and fans.

party line: A form of telephonic communication originating in the early twentieth century that enabled multiple users to be connected via the same circuit. Incoming calls for specific users could be differentiated through a distinctive ring assigned to each customer.

pasteurization: A process developed by Louis Pasteur by which food or drink is heated in order to slow microbial growth.

pastoral era: A utopian characterization of pre-industrialized life, where communities were composed primarily of locally based interactions in closely bounded groups.

penetration rate: The percentage of users of a technology in relation to non-adopters. It is used as a standard measure for comparing user groups when discussing the digital divide and global digital divide.

perpetual beta: The release of software to users before being complete, under the understanding that the software would continue to be improved based on user feedback.

persistence: Describes how data on social media and other digital tools is archived for later retrieval and use.

personal auditory space: The personal space that surrounds those immersed in listening to music (or other content) on their digital audio players (or MP3 players).

perspective effects: The development and application of techniques designed to provide the impression of depth and symmetry.

persuasion stage: A period when potential adopters actively seek out information about an innovation.

Pew Internet and American Life Project: An American organization that collects and analyzes data about American's use of the Internet and of related digital devices.

planum method: An archaeological technique primarily used in areas without clear stratification, involving lowering the site and recording the exposed surface.

political economic theory: A field of study that has different meanings. Often, the term refers to analysis based on Marxian theory. Other times, the term refers to studies influenced by the Chicago school.

Post-Fordist or post-industrial: A system of production and consumption that features greater technical and individual specificity in production and labour roles.

post-typography: A society that after having relied on print as a form of communication moves again toward a heavy reliance on orality.

primary orality: A term coined by Havelock to describe societies that rely only on oral forms of communication to transmit knowledge, information, and culture.

principle knowledge: An understanding of the mechanisms behind an innovation.

privacy: The right of individuals to choose what personal information they want to disclose to which audiences.

privacy paradox: How users disclose large amounts of personal information on the Internet, despite concerns about their privacy.

probes: Aphorisms consisting of ideas designed to inspire critical thinking about the media.

process of circular flow: A term Schumpeter uses to describe phases in the economy characterized by stability and a balance between demand and supply.

produsage: A term coined by Bruns describing the movement toward user-led areas of development, content creation, and collaboration of technologies and services, such as open-source software and interactive virtual communities.

prosumer: A term introduced by Toffler in 1980 to describe the merging of producers and consumers.

prosumption: Modes of production in which consumers perform a significant role in a product's design, development, and use.

public sphere: Public spaces where citizens can gather to share their perspectives and voice their opinions.

punishment-as-spectacle: The exertion of discipline in the Middle Ages through the use of techniques such as torture, decapitation, and burning.

quadrant method: An excavation technique used on circular sites, which involves digging four quadrants with baulks between them.

rational choice: A theory that models and attempts to understand social and economic behaviour.

rationalization: A new mode of operation in society, where social action is aimed at increasing efficiency and supporting a capitalist system of production rather than based on grounds of morality or tradition.

reflectionism: Ideas and procedures of using technology to mirror and confront surveillance.

regressive force: An influence that causes an entity, such as society, to degenerate or reverse from its established course, way of life, or mode of thinking.

relative advantage: The process by which the merits of an innovation are examined in relation to those of a previously existing idea, practice, or object.

relevant social group: Social constructivist concept pertaining to the group who gives meaning to an artifact.

remediation: The manner through which new media refashion prior media forms.

replicability: Capabilities provided online to reproduce content (text, pictures, and videos) and insert it in other contexts.

research and development (R&D): Characterized by a focus on the design, development, and implementation of innovations and inventions.

retro-analysis: The analysis of a situation after it has occurred; hence, it relies on memory, archives, and other documentation.

return on investment (ROI): Describes the economic gain to be made as a result of capital investment. It is usually measured as the ratio of monetary input to output.

revenge effect: An unintended effect occurring directly in an area of technological intervention.

reverse salient or **salience**: In the context of system theory, when a system grows unevenly, some components develop quickly, creating imbalances.

rhetorical closure: Occurs when instead of solving a technological problem, the meaning given to the problem is simply changed.

rich get richer hypothesis: That the Internet will primarily benefit those who are already connected and have large social networks instead of those who are isolated and have small social networks.

saturation: When a technology has reached complete penetration in a population.

science and technology studies (STS): An interdisciplinary field concerned with the study of how scientific and technological change intersect with society.

scientific management: Theories of management that attempt to use scientific methods to improve worker efficiency and productivity.

seamless Web of technology and society principle: The integration of political, economic, cultural, and social factors to explain how technological change occurs.

searchability: The ability to quickly locate information on people via the Web.

search history: Search and toolbar logs recorded by search engines and Internet browsers.

secondary orality: A term coined by Havelock to describe societies that move away from print and toward sound as a result of electronic media (including radio and television).

second order communication: A term Gershon used to convey the formality and nature of a medium.

sedentary societies: Small or large types of permanent settlement that are entirely immobile.

SGT STAR: A chatterbot used on the US Army website to provide visitors with information about army life.

side effect: An action or effect that occurs in an area unrelated to the one the technology was designed to function within.

simulation: Creating and enacting technologies that imitate the human mind or body within a computerized form.

skills divide: The gap between those who have the necessary skills to utilize information and communication technology and those who do not.

social affordance: Characterizes the relationship between the features of a system and the social characteristics of a group, facilitating or undermining particular social behaviours.

social after-effects: The social results of selecting or adopting one technology over another, which influence social relations and experiences.

social capital: The relationships forged through social interactions, networks, and connections, and the resources available through these relationships.

social change: Major change of structured social action or social structure taking place in a society, community, or social group.

social construction of technology (SCOT): The belief that technology is shaped by human needs and social factors. The cultural norms and values within a social system influence the construction, diffusion, and utilization of the technological product.

social determinism: The idea that social factors alone determine and shape uses of technology.

social informatics (SI): A concept based on social constructivist ideas that sees technology as emerging in dialectic with society.

social media: Web-based technologies, such as Facebook and YouTube, designed for social interaction; they often include a user profile, links between users, and virtual spaces for users to express their ideas.

social networks: The social structures formed by individuals' patterns of social interaction.

social network sites (SNS): Online websites, applications, or platforms centred on developing or maintaining virtual social relationships between individuals with shared interests, experiences, or connections.

social privacy: Protection of personal information from other people that we know, such as friends, co-workers, and family.

social structure: The arrangement of social organizations, institutions, and relationships within a society.

socio-technical perspective: The study of how social factors affect technological and scientific developments.

software cowboy: Programmers and developers of software who excel in their ability to write code and design software products.

sousveillance: The recording of events and activities via mobile technologies, such as cellphones, tablets, or wearable technologies.

space bias: Emerges in literate societies and favours imperialism and commerce. A space bias emerges because writing allows for the easy dissemination of ideas and information over vast distances.

spinoff companies: Companies that split from their parent company and develop their own business model based on ideas, products, or processes existent in the parent company.

s-shaped curve of adoption: The form that the adoption curve takes when we plot frequency of adoption rate on the y-axis and time on the x-axis.

stabilization: The point in which a relevant social group has assigned a very specific meaning and use to an artifact.

Stanford Industrial Park: An industrial park, today known as Stanford Research Park, located in Palo Alto, California, on land owned by Stanford University. Its aim is to bring together basic and applied research to develop cutting-edge products in the high-tech industry.

startup companies: Companies that start their operations based on a promising idea, often with limited capital and no revenue. The label *startup company* became well-known during the 1990s technology boom, when a new company surfaced every day.

Stone Age: A prehistoric era typically defined by the use of stone in the construction of tools.

strategic action: Planned action aimed at accomplishing specific goals with optimal results.

substantivism: A theory arguing that technology brings forth new social, political, and cultural systems, which it structures and controls.

supplement argument of the Internet: Argues that the Internet does not completely change social structure but, rather, becomes embedded in existing patterns of interaction.

surrounding awareness: A knowledge and understanding of ambient conditions outside of one's physical self.

surveillance: The monitoring of a person's behaviour or information, often with the use of technologies for recording and storing data.

synchronous forms of communication: Types of communication exchange—including Skype and Windows Live Messenger—that require the participation of each user in real time.

system: A configuration of components that are joined through linkages into a network.

systems theory: The field that studies social and biological systems that are self-regulating through, for example, feedback mechanisms.

Taylorism: A system of management developed by Taylor that aimed to make work more efficient and to increase productivity through the application of scientific standards and processes.

technical competence: The knowledge and/or skills required to demonstrate an understanding of the practices and tasks necessary to effectively utilize and master information and communication technologies.

technical or technocratic elite: A class of people who, according to social theorist Bell, emerge as the technocratic elite in a post-industrial information society due to their understanding of theoretical knowledge, technology, and information.

technique: A term used by Heidegger to describe how technologies constitute an object, a goal that the object fulfills, and a mechanism through which the goal is achieved.

technocracy: An organization, group, or collective comprising individuals from scientific, technological, and research-based fields, whose knowledge and technical expertise either influences or performs a defining role in making decisions regarding technical or scientific development.

technological determinism: A theory proposing that technology is the driving force in developing the structure of society and culture.

technological inequality: The gap between the haves and the have-nots, resulting in disadvantages for the latter.

technological regression: Rare cases where a technology is adopted and then later abandoned for social, political, cultural, or religious reasons. The social group then rejects the technology and returns to previously established practices and habits.

technological society: The pervasiveness and intrinsic centrality of technology in contemporary society.

technological utopianism: A belief that views technology as a largely positive force with the potential to create an idealized world through improved work conditions and lifestyles.

technology transfer: The flow of technological innovations from one country or region to another.

technopoles: Cities built within cities in order to produce the content, devices, and applications for the information society.

thanatechnology: A term Sofka uses to refer to the Internet; an instrument through which people can express emotional responses to death.

third places: Places that give people the opportunity to be social. Examples include coffee houses, taverns, restaurants, bars, libraries, and other locations that people visit routinely.

time bias: Innis's term to describe oral societies in which there is an emphasis on community and metaphysics. In these societies, change occurs slowly because the only transfer of culture, information, and knowledge is through oral means.

time study: A concept developed by Taylor involving the recording and detailing of a worker's movements to aid in the development of processes to improve efficiency.

Toronto School of Communication: A group of scholars working at the University of Toronto from about the 1940s to the 1970s, whose scholarship focused on the impact of media—including the alphabet, writing, television, and the radio—on society.

traditional technologies: Technologies of the pre-industrialized era, which are described as stationary and passive and having little effect on the natural environment.

transformation: The ability of separate entities in actor network theory (ANT) to convert into a single unit through a network.

transhumanism: A belief in the convergence of technology and the human body by using biotechnological components to augment or substitute existing physical human characteristics or parts with technological replacements.

transistor: A semiconductor that is used to route electronic signals.

triability: Provides potential adopters with an opportunity to reduce the uncertainty associated with new innovations by gathering evidence about its potential value and risks.

Turing test: A test developed by Turing to determine a machine's ability to showcase human intelligence.

Twitter Revolution: A term that describes the central role that Twitter played as an information and communication platform to organize and mobilize the citizens of a country to protest against their government.

uncertainty: A state of being unclear about the outcome of a particular innovation.

unintended effects: Consequences of technological invention or application, which are often only realized and understood after the technology has been used.

United Nations Development Program (UNDP): A global development program enacted by the United Nations to assist in solving global and national socio-economic problems related to education, poverty, and the environment.

unverifiability of power: Refers to the principle that one cannot determine whether or not one is being observed.

user-generated content: Content that is created, modified, and shared by end-users through Web 2.0 applications.

utopian perspective of technology: The belief that technology has a positive role in society and can be harnessed to create a more perfect world. Utopians embrace technology to improve efficiency and progress.

utopian perspective of the Internet: Views the Internet as leading toward positive social change, in particular by strengthening social relations and communities.

value-laden: A belief that technology is actively shaping or being shaped by culture, politics, or social values.

venture capital firms: Firms that specialize in investing in startup companies that have a high potential for growth.

virtual mourning: The use of Web-based media to mourn, commemorate, or provide support following the death of a family member, friend, or colleague.

virtual stranger loss: The means through which people mourn the death of individuals who may be strangers or whom they may not have met in person.

visibility of power: The principle that power is centrally displayed with the aim of intimidating.

voyeurism: The act of finding pleasure in secretly observing others engaged in private behaviours, such as undressing, sexual activity, etc.

wearable computing: Types of computers that are worn on the body. For instance, the WearComp is a type of wearable computer.

WearComp: A wearable computing system developed by Mann.

Web 2.0: Web-based applications that encourage and make possible interaction, information-sharing, and collaboration among users and producers.

WELL: Stands for Whole Earth 'Lectronic Link and is one of the first online communities to emerge on the Internet, based primarily in the San Francisco Bay Area.

Wheeler method: An archaeological excavation process that involves developing trenches in a series of rectangular boxes with balks in between them. The boxes assist in improving the organization of labour and comparing the origin of artifacts in relation to the adjacent layers of earth.

white-collar workers: Professional or educated workers whose job does not involve manual or physical labour.

wider context: Describes, in the context of social construction of technology (SCOT), how the value system of a social group influences the meaning given to and the use of a technology.

Wikimedia Foundation: A non-profit organization founded in 2003 by Wikipedia co-founder Jimmy Wales that operates various collaborative wiki-based projects on the Internet, such as Wikipedia.

Wikipedia: A popular editable and collaborative non-profit, Web-based encyclopedia administered by the Wikimedia Foundation.

wikis: Websites using simplified markup language that allows users to collaborate on projects and to edit or add text.

write community into being: Through friending—i.e., making friends on social network sites—users establish the boundaries of their virtual community.

write themselves into being: The idea that through online participation a user's profile can represent and his or her virtual self.

XO laptop: A sturdy and low-cost laptop that employs open-source software designed specifically for the One Laptop per Child program.

Notes

Chapter 1

1. This reference to *Star Trek* was used previously in Hogan and Quan-Haase with reference to the rapid changes occurring in the realm of social media (2010).

2. The definition of *artifact* was taken from the *Concise Oxford Dictionary of Archaeology* and focuses primarily on the distinction between objects created by humans and objects occurring naturally. (*The Concise Oxford Dictionary of Archaeology*. Timothy Darvill. Oxford University Press, 2008. Oxford Reference Online. Oxford University Press.)

3. The term *normalized* was first used by Pacey (1983) in his ideas of how technology becomes embedded in a culture. The term has been used as well in the context of the debate on the impact of the Internet on society. For instance, Herring uses the term in her article on how computer-mediated communication has become ordinary in society (2004).

4. For a similar list describing the challenges associated with studying the effects of the Internet on society, see Quan-Haase and Wellman (2004).

5. Information on the cyborg models comes from the Wikipedia entry on Terminator (character). http://en.wikipedia.org/wiki/Terminator_(character) Retrieved 4 June 2012.

Chapter 2

1. Information from the Stone Age comes from the Southern Africa entry from the Encyclopaedia Britannica: Academic Edition Online, www.britannica.com/EBchecked/topic/556618/Southern-Africa?cameFromBol=true. Retrieved 13 October 2010.

2. Further information is available on the CBC McLuhan archives. Retrieved from http://archives.cbc.ca/arts_entertainment/media/topics/342-1818/.

3. A review of Pertierra et al.'s book *Txt-ing Selves* (2003) can be found in Quan-Haase, A. (2004). Book Review of Pertierra, R., Ugarte, E.F., Pingol, A., Hernandez, J., & Dacanay, N.L. (2003). *Txt-ing selves: Cellphones and Philippine modernity*. Manila: De La Salle University Press. *Canadian Journal of Sociology*.

Chapter 3

1. See Sorensen, C. (2010, July 29). Toyota's latest repairs. *Maclean's*. Retrieved from www2.macleans.ca/2010/07/29/toyotas-latest-repairs/. See also Russell, R. (2011, July 26). An inconvenient truth about Toyota. *The Globe and Mail*. Retrieved from http://m.theglobeandmail.com/globe-drive/car-tips/safety/an-inconvenient-truth-about-toyota/article2104879/?service=mobile

2. For a detailed review, see Kling (1999), and for examples of such studies see the special issue of the *Journal of the American Society for Information Science* (Kling, Rosenbaum, & Hert, 1998).

Chapter 4

1. The GDP measures a nation's standard of living as it is an indicator of the value of goods and services produced, usually within a one-year period, and it was first introduced by Simon Kuznets in 1934. It is usually calculated as GDP = consumption + gross investments + government spending + net exports.

Chapter 5

1. *Scholars Portal* is a database that contains a comprehensive selection of e-journals and is available to all universities in Ontario. See http://journals1.scholarsportal.info/discuss.xqy.

2. This information comes from Arvind Singhal's speech for the introduction of the award ceremonies for Professor Everett M. Rogers when he was named the 47th Annual Research Lecturer at the University of New Mexico, Ohio University, 24 April, 2002. Retrieved 5 July, 2010, from www.insna.org/PDF/Connections/v26/2005_I-2-2.pdf.

3. Cultures are constantly changing and, as a result, the values and meanings given to technologies also constantly need updating.

4. This phenomenon has recently been observed with the BlackBerry, where users switched to other devices but then realized that they preferred the features offered by the BlackBerry and so switched back.

Chapter 6

1. In 2011, the number of Canadian Facebook users was estimated at 16.5 million (www.insidefacebook.com/2011/06/12/facebook-sees-big-traffic-drops-in-us-and-canada-as-it-nears-700-million-users-worldwide/).

2. Often, software products are updated as new features are developed and integrated into the system.

Chapter 7

1. Data come from the 1995 Household Facilities and Equipment Survey and exclude residents from the Yukon and Northwest Territories, Indian reserves and Crown land and institutions (Dickinson & Sciadas, 1996). At the time of the survey, owning a TV without cable was still fairly common.

2. Further information on E-Rate and the source of the quote can be found at www.fcc.gov/learnnet/ (retrieved 2 May, 2011).

3. These numbers are per 100 inhabitants.

4. In the categorization used by the ITU report, OECD nations constitute the most developed nations and include North America and Europe. Developing nations or industrializing nations are less developed, but moving toward heavy industrialization and include India and China. The least developed countries are the poorest and least industrialized and include Latin America and African nations.

5. This would mean that an individual living in an OECD country would be 9.8 times more likely than a person living in an developing country to have a fixed line.

6. As an example, the numbers of the Democratic Republic of the Congo show that the ratio of cellphones to fixed telephone lines in 2009 is 254 to 1.

Chapter 9

1. The sources in this chapter are from the CBC documentary *Human Journey* (2010), which documents the spread of the human race across the five continents. The series brings together new scientific evidence from scholars that shows the path believed to have been taken by humans in their travels throughout the continents. The five-part series can be watched online at www.cbc.ca/documentaries/passionateeyeshowcase/2010/humanjourney/.

2. The concept of friendship is also closely linked to the distinction made by Granovetter (1973) between weak and strong ties. Strong ties are with individuals who one trusts, has frequent contact with, and who provide social support in times of need.

3. This passage is from Nora Jones's show *Spark*. You can listen to the full show at www.cbc.ca/spark/2010/04/spark-111-april-25-27-2010.

Chapter 10

1. Sting describes in an interview how the separation from his wife motivated him to write the song *Every Breath You Take*. For a detailed analysis of the depiction of surveillance practices in popular culture, see G. Marx (1996). Electric eye in the sky: Some reflections on the new surveillance and popular culture. In D. Lyon & E. Zureik (Eds.), *Computers, surveillance, and privacy* (pp. 193–233). Minneapolis: University of Minnesota Press.

2. In a normalized system, there is little room for a person to be different or unique. Nonetheless, fitting into the system is not always a measure of success. For instance, Bill Gates, Mark Zuckerberg, and Steve Jobs are all extremely successful entrepreneurs in the technology sector who dropped out of the school system.

3. Also see the discussion in Chapter 9 concerning the four characteristics of social network sites identified by boyd in her research on Friendster (2006).

Chapter 11

1. Many AKs in use were variants such as AK-74, AKSU (Bin Laden's little gun), AKM, as well as foreign copies with different names (Chinese Type 56, Yugoslav M90). Powers other than the USSR sold the AK-47, such as China.

2. The weapon was designed based on the impetus of the German invasion; however, by the time trials and early production were complete, the AK was several years too late to influence the war. The two primary Soviet arms developments that changed the nature of WWII were the invention of the T-34 tank (fast, well armoured, and easy to mass-produce) and the PPSH-41 submachine gun (high rate of fire, cheap, and reliable).

3. This is particularly true in poorer regions of the world, which barter materials rather than purchase goods in cash. Moses Naim points out that in 1986 an individual in Kenya would have had to trade 15 cows for an AK-47; in 2005, it would have cost the same individual only 4 cows (Naim & Myers, 2005).

4. Even though they were referred to as the Fukushima 50 Heroes, there were many more workers at the plant. Some news reports estimate that up to 700 workers took shifts during the crisis.

References

Abrams, J. J. (2009). *Star Trek* [motion picture]. Retrieved from www.imdb.com/title/tt0796366/quotes

Adams, P. (2011). Why I left Google. What happened to my book. What I work on at Facebook [Blog post]. *Think Outside In.* Retrieved from www.thinkoutsidein.com/blog/2011/07/why-i-left-google-what-happened-to-my-book-what-i-work-on-at-facebook/

Andrejevic, M. (2004). *Reality TV: The work of being watched.* Lanham, MD: Rowman & Littlefield.

Back, M. D., Stopfer, J. M., Vazire, S., Gaddis, S., Schmukle, S. C., Egloff, B., et al. (2010). Facebook profiles reflect actual personality, not self-idealization. *Psychological Science China, 21*(3), 371–374.

Bánáthy, B. H. (1997, February 28). A taste of systemics. Retrieved from www.newciv.org/ISSS_Primer/asem04bb.html

Bandura, A. (1973). *Aggression: A social learning analysis.* Englewood Cliffs, NJ: Prentice-Hall.

Barnes, S. B. (2006). A privacy paradox: Social networking in the United States. *First Monday, 11*(9). Retrieved from http://firstmonday.org/htbin/cgiwrap/bin/ojs/index.php/fm/article/view/1394/1312

Barrett, M., Grant, D., & Wailes, N. (2006). ICT and organizational change. *The Journal of Applied Behavioral Science, 42*(1), 6–22.

Basalla, G. (1988). *The evolution of technology.* Cambridge: Cambridge University Press.

Batson, A. (2010). China's GDP: Still number three. *The Wall Street Journal.* Retrieved from http://blogs.wsj.com/chinarealtime/2010/07/02/chinas-gdp-still-number-three/

Bauchspies, W. K., Restivo, S. P., & Croissant, J. (2006). *Science, technology, and society: A sociological approach.* Malden, MA: Blackwell.

Baudrillard, J. (2005). *The system of objects.* London: Verso.

———, & Gane, M. (1993). *Baudrillard live: Selected interviews.* London: Routledge.

Bauwens, M. (2005, November 4). P2P and human evolution: Placing peer to peer theory in an integral framework. Retrieved from http://integralvisioning.org/article.php?story=p2ptheory1

Baym, N. K. (2010). *Personal connections in the digital age.* Cambridge, UK: Polity Press.

BBC. (2006, November 27). Star Wars kid is top viral video. *BBC News.* Retrieved from http://news.bbc.co.uk/2/hi/entertainment/6187554.stm

Beckett, A. (2010, July 12). Inside the Bill and Melinda Gates foundation. *The Guardian.* Retrieved from www.guardian.co.uk/world/2010/jul/12/bill-and-melinda-gates-foundation

Bell, D. (1973). *The coming of a post-industrial society: A venture in social forecasting.* New York: Basic Books.

Benkler, Y. (2006). *The wealth of networks: How social production transforms markets and freedom.* New Haven, CT: Yale University Press.

Berg, M. (1994). *The age of manufactures, 1700–1820: Industry, innovation and work in Britain* (2nd ed.). New York: Routledge.

Bertot, J. C., Jaeger, P. T., McClure, C. R., Wright, C. B., & Jensen, E. (2009). Public libraries and the Internet 2008–2009: Issues, implications, and challenges. *First Monday, 14*(11). Retrieved from www.uic.edu/htbin/cgiwrap/bin/ojs/index.php/fm/article/view/2700/2351

Bijker, W. E. (2009). Social construction of technology. In J.-K. B. Olsen, S. A. Pedersen & V. F. Hendricks (Eds.), *A companion to the philosophy of technology* (pp. 88–94). Malden, MA: Wiley-Blackwell.

———, Hughes, T. P., & Pinch, T. (1999). General introduction. In W. E. Bijker, T. P. Hughes & T. Pinch (Eds.), *The social construction of technological systems: New directions in the sociology and history of technology* (pp. 1–6). Cambridge, MA: MIT Press.

Bill and Melinda Gates Foundation. (2011). About the foundation. Retrieved from www.gatesfoundation.org/about/Pages/values.aspx

Bilton, N., & Rusli, E. (2012). From founders to decorators, Facebook riches. Retrieved 15 February 2012, from www.nytimes.com/2012/02/02/technology/for-founders-to-decorators-facebook-riches.html?pagewanted=all

Binfield, K. (Ed.). (2004). *Writings of the Luddites.* Baltimore, MD: Johns Hopkins University Press.

Boehm, C. (1993). Egalitarian behavior and reverse dominance hierarchy. *Current Anthropology, 34,* 227–254.

——— (1999). *Hierarchy in the forest: The evolution of egalitarian behavior.* Cambridge, MA: Harvard University Press.

——— (2000). The origin of morality as social control. In L. D. Katz (Ed.), *Evolutionary origins of morality: Cross-disciplinary perspectives.* Bowling Green, OH: Imprint Academic.

Bogart, L. (1972). *The age of television; a study of viewing habits and the impact of television on American life* (3rd ed.). New York: F. Ungar.

Bolter, J. D., & Grusin, R. (1999). *Remediation: Understanding new media.* Cambridge, MA: MIT Press.

Bonabeau, E. (2009). Decisions 2.0: The power of collective intelligence. *MIT Sloan Management Review, 50*(2), 45–52.

Boradkar, P. (2006). 10,000 songs in your pocket: The iPod® as a transportable environment. In R. Kronenburg & F. Klassen (Eds.), *Transportable environments 3* (pp. 21–30). Abingdon: Taylor & Francis.

Boreham, P., Parker, R., Thompson, P., & Hall, R. (2008). *New technology @ work.* New York: Routledge.

Borgatti, S. P. (1998). A SOCNET discussion on the origins of the term social capital. *Connections, 21*(2), 37–46.

Borgmann, A. (1984). Technology and democracy. In P. T. Durbin (Ed.), *Research in philosophy & technology : Vol. 7* (pp. 211–228). Greenwich, CT: Jai Press.

Borsook, P. (2000). *Cyberselfish: A critical romp through the terribly libertarian culture of high tech.* New York: Public Affairs.

boyd, d. (2006). Friends, Friendsters, and top 8: Writing community into being on social network sites. *First Monday, 11*(12). Retrieved from http://firstmonday.org/htbin/cgiwrap/bin/ojs/index.php/fm/article/view/1418/1336

——— (2007). Why youth (heart) social network sites: The role of networked publics in teenage social life. In D. Buckingham (Ed.), *The MacArthur foundation series on digital learning—youth, identity and digital media volume* (pp. 119–142). Cambridge, MA: MIT Press.

———, & Heer, J. (2006). *Profiles as conversation: Networked identity performance on Friendster.* Proceedings of the Hawai'i International Conference on System Sciences (HICSS-39), Persistent Conversation Track. IEEE Computer Society, Kauai, HI. Retrieved from www.danah.org/papers/HICSS2006.pdf

Bradner, E. (2001). Social affordances of computer-mediated communication technology: Understanding adoption. In *CHI '01 extended abstracts on human factors in computing* pp. 67–8. New York: ACM.

———, Kellogg, W. A., & Erickson, T. (1999, 12–16 September). *Social affordances of Babble: A field study of chat in the workplace.* ECSCW, 99, the Sixth European Conference on Computer Supported Cooperative Work, Copenhagen, Denmark.

Braverman, H. (1974). *Labor and monopoly capital: The degradation of work in the twentieth century.* New York: Monthly Review Press.

Briggs, A., & Burke, P. (2009). *A social history of the media: From Gutenberg to the Internet* (3rd ed.). Cambridge, UK: Polity.

Bruns, A. (2008). *Blogs, Wikipedia, Second Life, and beyond: From production to produsage.* New York: Peter Lang.

Bull, M. (2008). *Sound moves: Ipod culture and urban experience.* New York: Routledge.

Bynum, W. F., & Porter, R. (Eds.). (2006). *The Oxford dictionary of scientific quotations.* Oxford University Press.

Callon, M. (1987). Society in the making: The study of technology as a tool for sociological analysis. In W. E. Bijker, T. P. Hughes & T. Pinch (Eds.), *The social construction of technological systems: New directions in the sociology and history of technology* (pp. 83–103). Cambridge, MA: MIT Press.

———, & Law, J. (1997). After the individual in society: Lessons on collectivity from science, technology and society. *Canadian Journal of Sociology, 22*(2), 165–182.

CANARIE. (2010, August 10). About CANARIE. Retrieved from www.canarie.ca/en/about/aboutus

Carmel, E. (1995). Cycle-time in packaged software firms. *Journal of Product Innovation Management, 12*(2), 110–123.

———, & Sawyer, S. (1998). Packaged software development teams: What makes them different? *Information, Technology & People, 11*(1), 7–19.

Carroll, B., & Landry, K. (2010). Logging on and letting out: Using online social networks to grieve and to mourn. *Bulletin of Science, Technology and Society, 30*(5), 377–386. Retrieved from http://bst.sagepub.com/content/30/5/341.full.pdf+html

Castells, M. (1996). *The rise of the network society.* Cambridge, MA: Blackwell Publishers.

———— (2001). *The Internet galaxy: Reflections on the Internet, business, and society.* Oxford: Oxford University Press.

————, & Hall, P. G. (1994). *Technopoles of the world: The making of twenty-first-century industrial complexes.* London: Routledge.

Ceruzzi, P. E. (2005). Moore's law and technological determinism: Reflections on the history of technology. *Technology and Culture, 46*(3), 584–593.

Chen, W., Boase, J., & Wellman, B. (2002). The global villagers: Comparing Internet users and uses around the world. In B. Wellman & C. Haythornthwaite (Eds.), *The Internet in everyday life* (pp. 74–113). Oxford: Blackwell.

Chesterman, S. (2011). *One nation under surveillance: A new social contract to defend freedom without sacrificing liberty.* Oxford: Oxford University Press.

Clinton, H. R. (2011, February 15). Secretary Clinton on Internet rights and wrongs: Choices & challenges in a networked world. [Video clip.] Retrieved from http://geneva.usmission.gov/2011/02/16/internet-rights-and-wrongs/

Clynes, M. E., & Kline, N. S. (1960). Cyborgs and space. *Austronautics* (September), 26–27 & 74–76.

Coleman, J. S., Katz, E., & Menzel, H. (1966). *Medical innovation: A diffusion study.* New York: Bobbs Merrill.

Connolly, K. (2002, July 30). "I wish I'd made a lawnmower." *The Guardian.* Retrieved from www.guardian.co.uk/world/2002/jul/30/russia.kateconnolly

Constantine, L. L. (1995). *Constantine on peopleware.* Englewood Cliffs, NJ: Prentice Hall.

Costa, D. L., & Kahn, M. E. (2001). *Understanding the decline in social capital, 1952–1998.* Retrieved from UCLA website: www.econ.ucla.edu/costa/scapital8.pdf

Crowley, D. J., & Heyer, P. (2011). *Communication in history: Technology, culture, society* (6th ed.). Boston: Allyn & Bacon.

Cullen, R. (2001). Addressing the digital divide. *Online Information Review, 25*(5), 311–320.

Cutcliffe, S. H., & Mitcham, C. (2001). Introduction: The visionary challenges of STS. In S. H. Cutcliffe & C. Mitcham (Eds.), *Visions of STS: Counterpoints in science, technology, and society studies* (pp. 1–7). Albany, NY: State University of New York Press.

David, P. A. (1985). Clio and the economics of QWERTY. *American Economic Review 75*(2), 332–337.

Day, P., & Schuler, D. (2006). Community practice in the networked society: Pathways towards civic intelligence. In P. A. Purcell (Ed.), *Networked neighbourhoods: The connected community in context* (pp. 19–46). London: Springer.

Dedehayir, O. (2009). Bibliometric study of the reverse salient concept. *Journal of Industiral Engineering and Management, 2*(3), 569–591. Retrieved from http://upcommons.upc.edu/revistes/bitstream/2099/8490/1/dedehayir.pdf

————, & Mäkinen, S. J. (2008). Dynamics of reverse salience as technological performance gap: An empirical study of the personal computer technology system. *Journal of Technology Management & Innovation, 3*(4). Retrieved from www.scielo.cl/scielo.php?pid=S0718-27242008000100006&script=sci_arttext

Dewdney, C. (1998). *Last flesh: Life in the transhuman era.* Toronto: HarperCollins.

———— (2001, December 15). First Mann into cyborgspace. *The Globe and Mail,* pp. D6–7.

Diamond, J. M. (1997). *Guns, germs, and steel: The fates of human societies.* New York: Norton.

Dickinson, P., & Sciadas, G. (1996). *Access to the information highway.* Services Science and Technology Division. Retrieved from www.statcan.gc.ca/pub/63f0002x/63f0002x1996009-eng.pdf

Dinwiddy, J. (1979). Luddism and politics in the northern counties. *Social History, 4*(1), 33–63.

Donath, J., & boyd, d. (2004). Public displays of connection. *BT Technology Journal, 22*(4), 71–82.

Dubé, L. (1998). Teams in packaged software development: The software corp. experience. *Information, Technology & People, 11*(1), 36–61.

Dyer-Witheford, N. (1999). *Cyber-Marx: Cycles and circuits of struggle in high-technology capitalism.* Urbana, IL: University of Illinois Press.

———— (2001). Empire, immaterial labor, the new combinations, and the global worker. *Rethinking Marxism, 13*(3/4), 70–80.

Edmunds, S. (1991). From Schoeffer to Vérard: Concerning the scribes who became printers. In S. Hindman (Ed.), *Printing the written word: The social history of books, circa 1450–1520* (pp. 21–40). Ithaca: Cornell University Press.

Education and Library Networks Coalition. (2007). *E-rate: 10 years of connecting kids and community.* National Coalition for Technology in Education and Training.

Retrieved from www.edlinc.org/pdf/ NCTETReport_212.pdf

Edwards, R. (1979). *Contested terrain: The transformation of the workplace in the twentieth century.* New York: Basic Books.

Eisenstein, E. L. (1979). *The printing press as an agent of change: Communications and cultural transformations in early modern Europe.* Cambridge: Cambridge University Press.

Eisler, P., Schmit, J., & Leinwand, D. (2011, March 15). Japan's nuke threat "a wake-up call" for the U.S. *USA Today.* Retrieved from www.usatoday.com/money/industries/energy/2011-03-14-japan-nuclear-crisis-a-wake-up-call_N.htm

Elliott, J. E. (2004). Introduction to the transaction edition. In J. A. Schumpeter (Ed.), *The theory of economic development: An inquiry into profits, capital, credit, interest, and the business cycle* (pp. vii–lix). Cambridge: Harvard University Press.

Ellison, N. B., Steinfield, C., & Lampe, C. (2007). The benefits of Facebook "Friends:" Social capital and college students' use of online social network sites. *Journal of Computer-Mediated Communication, 12*(4), article 1. Retrieved from http://jcmc.indiana.edu/vol12/issue4/ellison.html

Ellul, J. (1964). *The technological society* (1st American ed.). New York: Knopf.

——— (1989). *What I believe.* Grand Rapids, MI: W.B. Eerdmans.

———, & Vanderburg, W. H. (1981). *Perspectives on our age: Jacques Ellul speaks on his life and work.* New York: Seabury Press.

Epstein, D., Nisbet, E. C., & Gillespie, T. (2011). Who's responsible for the digital divide? Public perceptions and policy implications. *The Information Society, 27*(2), 92–104.

Facebook. (2011). Facebook statistics. Retrieved from www.facebook.com/press/info.php?statistics

Fallows, D. (2005). *Search engine users.* Retrieved from Pew Internet and American Life Project website: www.pewinternet.org/Reports/2005/Search-Engine-Users.aspx

Farley, T. (2005). Mobile telephone history. *Telektronikk, 3*(4), 22–34. Retrieved from www.privateline.com/archive/TelenorPage_022-034.pdf

Febvre, L. P. V., & Martin, H.-J. (1997). *The coming of the book: The impact of printing, 1450–1800.* London: Verso.

Federal Communications Commission (FCC). (2004, May 2). E-rate. Retrieved from www.fcc.gov/learnnet/

Feenberg, A. (1982). Technology and the idea of progress. In P. T. Durbin (Ed.), *Research in philosophy & technology* (Vol. 5, pp. 15–21). Greenwich, CT: Jai Press.

——— (1991). *Critical theory of technology.* New York: Oxford University Press.

——— (1999). *Questioning technology.* New York: Routledge.

Feist, R., Beauvais, C., & Shukla, R. (2010). Introduction. In R. Feist, C. Beauvais & R. Shukla (Eds.), *Technology and the changing face of humanity* (pp. 1–21). Ottawa: University of Ottawa Press.

Ferguson, E. S. (1992). *Engineering and the mind's eye.* Cambridge, MA: MIT Press.

Fischer, C. S. (1982). *To dwell among friends.* Berkeley, CA: University of California Press.

——— (1992). *America calling: A social history of the telephone to 1940.* Berkeley, CA: University of California Press.

Flew, T. (2008). *New media: An introduction.* Melbourne, Australia: Oxford University Press.

Florida, R. (2001, December 5). *The creative class.* Lecture presented at Rotman School of Management, University of Toronto, Toronto, ON.

Foucault, M. (1995). *Discipline and punish: The birth of the prison* (1st American ed.). New York: Vintage Books.

Fowler, W. S. (1962). *The development of scientific method.* Oxford: Pergamon Press.

Francis, R. D. (2009). *The technological imperative in Canada: An intellectual history.* Vancouver: UBC Press.

Franklin, U. M. (1992). *The real world of technology.* Concord, ON: House of Anansi Press.

Franz, C. R., Roby, D., & Koeblitz, R. R. (1986). User response to an online information system: A field experiment. *MIS Quarterly, 10*(1), 29–42.

Fuchs, C. (2008). *Internet and society: Social theory in the information age.* New York: Routledge.

——— (2010). Labor in informational capitalism and on the Internet. *The Information Society, 26*(3), 179–196.

Füssel, S. (2005). *Gutenberg and the impact of printing* (1st English ed.). Aldershot, UK: Ashgate Publishing.

Gasher, M., Skinner, D. & Lorimer, R. (2012). *Mass communication in Canada* (7th ed.). Don Mills, ON: Oxford University Press.

Gershon, I. (2010). *The breakup 2.0: Disconnecting over new media.* Ithaca, NY: Cornell University Press.

Goffman, E. (1959). *The presentation of self in everyday life.* Garden City, NY: Doubleday.

Granovetter, M. S. (1973). The strength of weak ties. *American Journal of Sociology, 78,* 1360–1380.

Grant, G. P. (2002). Philosophy in the mass age. In A. Davis (Ed.), *Collected works of George Grant: 1951–1959* (Vol. 2). Toronto: University of Toronto Press.

Grant, G. P. (1986). *Technology and justice.* Toronto: House of Anansi Press.

Green, L. (2002). *Communication, technology and society.* London: SAGE.

Greenberg, A. (2008). The privacy paradox. *Forbes.* Retrieved from www.forbes.com/2008/02/15/search-privacy-ask-tech-security-cx_ag_0215search.html

Grint, K., & Woolgar, S. (1997). *The machine at work: Technology, work and organization.* Malden, MA: Blackwell.

Gripsrud, J., Moe, H., & Splichal, S. (2010). *The digital public sphere: Challenges for media policy.* GÅteborg: Nordicom.

Gross, R., & Acquisti, A. (2005). *Information revelation and privacy in online social networks.* Proceedings of the 2005 ACM workshop on privacy in the electronic society, New York. Retrieved from www.heinz.cmu.edu/~acquisti/papers/privacy-facebook-gross-acquisti.pdf

Grossberg, L. (1996). On postmodernism and articulation: An interview with Stuart Hall. In D. Morley & K.-H. Chen (Eds.), *Stuart Hall: Critical dialogues in cultural studies* (pp. 131–150). London: Routledge.

Guest, A. M., & Wierzbicki, S. K. (1999). Social ties at the neighborhood level: Two decades of GSS evidence. *Urban Affairs Review, 35*(1), 92–111.

Gurstein, M. (2007). *What is community informatics (and why does it matter)?* Milano: Polimetrica.

Gutting, G. (2010). Michel Foucault. *Stanford Encyclopedia of Philosophy.* Retrieved from http://plato.stanford.edu/entries/foucault/

Ha, L., & James, E. L. (1998). Interactivity re-examined: A baseline analysis of early business websites. *Journal of Broadcasting & Electronic Media, 42*(4), 457–474.

Haas, C. (1996). *Writing technology: Studies on the materiality of literacy.* Mahwah, NJ: Lawrence Erlbaum Associates.

Habermas, J. (1984). *The theory of communicative action: Reason and rationalization of society.* Boston: Beacon Press.

Haigh, T. (2011). The history of information technology. *Annual Review of Information Science and Technology 45,* 431–487.

Hake, S. (2002). *German national cinema.* New York: Routledge.

Halverson, J. (1992). Havelock on Greek orality and literacy. *Journal of the History of Ideas, 53*(1), 148–163.

Hanks, C. (Ed.). (2010). *Technology and values.* Malden, MA: Wiley-Blackwell.

Hargittai, E. (2002). Second-level digital divide: Differences in people's online skills. *First Monday, 7*(4). Retrieved from http://firstmonday.org/htbin/cgiwrap/bin/ojs/index.php/fm/article/view/942/864

Hartley, M. (2009, May 12). How the iPod changed everything. *The Globe and Mail.* Retrieved from www.theglobeandmail.com/news/technology/download-decade/how-the-ipod-changed-everything/article1133329/

Hauser, G. A. (1998). Vernacular dialogue and the rhetoricality of public opinion. *Communication Monographs, 65*(2), 83–107.

Havelock, E. A. (1963). *Preface to Plato.* Cambridge: Harvard University Press.

Haythornthwaite, C. (2005). Social networks and Internet connectivity effects. *Information, Communication & Society, 8*(2), 125–147.

Hearn, A. (2006). "John, a 20–year-old Boston native with a great sense of humour": On the spectacularization of the "self" and the incorporation of identity in the age of reality television. *International Journal of Media and Cultural Politics, 2*(2), 131–147. doi: 10.1386/macp.2.2.131/1.

Heidegger, M. (2010). The question concerning technology. In C. Hanks (Ed.), *Technology and values: Essential reader* (pp. 99–114). Malden, MA: Wiley-Blackwell.

Henton, D., Melville, J., Grose, T., Furrell, T., Halter, G., Harutyunyan, A., et al. (2011). Index of Silicon Valley. Retrieved from www.jointventure.org/images/stories/pdf/The%20Index%20of%20Silicon%20Valley%202011.pdf

Herring, S. C. (2004). Slouching toward the ordinary: Current trends in computer-mediated communication. *New Media and Society, 6*(1), 26–36.

Hershbach, D. (1995). Technology as knowledge: Implications for instruction. *Journal of Technology Education, 7*(1), 31–42.

Hill, C. T. (1989). Technology and international competitiveness: Metaphor for progress. In S. L. Goldman (Ed.), *Science, technology, and social progress* (pp. 33–47). Cranbury, NJ: Associated University Presses.

Hippel, E. v. (2005). *Democratizing innovation.* Cambridge, MA: MIT Press.

Hobsbawn, E. J. (1952). The machine breakers. *Past and Present, 1*(1), 57–70.

Hogan, B. (2010). The presentation of self in the age of social media: Distinguishing performances and exhibitions online. *Bulletin of Science, Technology & Society, 30*(6), 377–386. Retrieved from http://bst.sagepub.com/content/30/6/377.full.pdf+html

Hogan, B., & Quan-Haase, A. (2010). Persistence and change in social media: A framework of social practice. *Bulletin of Science, Technology and Society, 30*(5), 309–315. Retrieved from http://bst.sagepub.com/content/30/5/309.full.pdf+html

Holzmann, G. J., & Pehrson, B. (1995). *The early history of data networks.* Los Alamitos, CA: IEEE Computer Society Press.

Horrigan, J. B. (2008). *Home broadband adoption 2008.* Retrieved from Pew Internet and American Life Project website: www.pewinternet.org/~/media//Files/Reports/2008/PIP_Broadband_2008.pdf

Horton, D., & Wohl, R. R. (1956). Mass communication and para-social interaction: Observations on intimacy at a distance. *Psychiatry 19*, 215–229.

Hounshell, D. A. (1985). *From the American system to mass production, 1800–1932: The development of manufacturing technology in the United States.* Baltimore: Johns Hopkins University Press.

Howard, P. E. N., Rainie, L., & Jones, S. (2002). Days and nights on the Internet: The impact of a diffusing technology. In B. Wellman & C. Haythornthwaite (Eds.), *Internet and everyday life* (pp. 45–73). Oxford: Blackwell Publishers.

——— (2011). *The digital origins of dictatorship and democracy: Information technology and political Islam.* Oxford: Oxford University Press.

Hughes, T. P. (1983). *Networks of power: Electrification in Western society, 1880–1930.* Baltimore: Johns Hopkins University Press.

Huws, U. (2003). *The making of a cybertariat: Virtual work in a real world.* New York: Monthly Review Press.

Ihde, D. (2010). *Heidegger's technologies: Postphenomenological perspectives.* New York: Fordham University Press.

Innis, H. A. (1951). *The bias of communication.* Toronto: University of Toronto Press.

International Telecommunication Union. (2009). *Mobile cellular subscriptions.* ITU World Telecommunication/ICT Indicators Database. Retrieved from www.itu.int/ITU-D/icteye/Reporting/ShowReportFrame.aspx?ReportName=/WTI/CellularSubscribersPublic&ReportFormat=HTML4.0&RP_intYear=2009&RP_intLanguageID=1&RP_bitLiveData=False

Jacobs, J. (1961). *The death and life of great American cities.* NY: Random House.

Jarvenpaa, S. L., & Ives, B. (1994). The global network organization of the future: Information management opportunities and challenges. *Journal of Management Information Systems, 10*(4), 25–57.

Jarvis, J. (2009). *What would Google do?* (1st ed.). New York, NY: Collins Business.

Jobs, S. To all iPhone customers. Retrieved from www.apple.com/hotnews/openiphoneletter/

Johns, A. (1998). *The nature of the book: Print and knowledge in the making.* Chicago, IL: University of Chicago Press.

Jonas, H. (1984). *The imperative of responsibility: In search of an ethics for the technological age.* Chicago: University of Chicago Press.

——— (2003). Toward a philosophy of technology. In R. C. Scharff & V. Dusek (Eds.), *Philosophy of technology: The technological condition: An anthology* (pp. 191–204). Malden, MA: Blackwell Publishers.

Jones, S. E. (2006). *Against technology: From the Luddites to neo-Luddism.* New York: Routledge.

Kahaner, L. (2007). *AK-47: The weapon that changed the face of war.* Hoboken, NJ: John Wiley & Sons.

Kaplan, A. M., & Haenlein, M. (2010). Users of the world, unite! The challenges and opportunities of social media. *Business Horizons, 53*(1), 59–68.

Katz, J. E., & Rice, R. E. (2002). Syntopia: Access, civic involvement, and social interaction on the net. In B. Wellman & C. Haythornthwaite (Eds.), *The Internet in everyday life.* Oxford: Blackwell Publishers.

Kim, S. H. (2008, December 31). Max Weber. *Stanford Encyclopedia of Philosophy.* Retrieved from http://plato.stanford.edu/archives/fall2008/entries/weber/

Kirby, P. (2002). *The Celtic Tiger in distress: Growth with inequality in Ireland.* New York: Palgrave.

——— (2010). *Celtic Tiger in collapse.* New York: Palgrave.

Kirkpatrick, D. (2010). *The Facebook effect: The inside story of the company that is connecting the world.* New York: Simon & Schuster.

Klemens, G. (2010). *The cellphone: The history and technology of the gadget that changed the world.* Jefferson, N.C.: McFarland & Co.

Kling, R. (1999). Learning about information technologies and social change: The contribution of social informatics. *The Information Society, 16,* 217–232.

———, Rosenbaum, H., & Hert, C. (1998). Social informatics in information science: An introduction. *Journal of the American Society of Information Science, 49*(12), 1047–1052.

Kraemer, K. L., Dedrick, J., & Sharma, P. (2009). One laptop per child: Vision versus reality. *Communications of the ACM, 52*(6), 66–73.

Kraut, R. E., Kiesler, S., Boneva, B., Cummings, J., Helgeson, V., & Crawford, A. (2002). Internet paradox revisited. *Journal of Social Issues, 58*(1), 49–74.

Krishnan, M. S. (1998). The role of team factors in software cost and quality: An empirical analysis. *Information, Technology & People, 11*(1), 20–35.

Latour, B. (1987). *Science in action: How to follow scientists and engineers through society.* Milton Keynes: Open University Press.

Lampe, C., Ellison, N. B., & Steinfield, C. (2007). *A familiar face(book): Profile elements as signals in an online social network.* CHI 2007 Online Representation of Self, San Jose, CA. Retrieved from www.msu.edu/~steinfie/CHI_manuscript.pdf

Latour, B. (1987). *Science in action: How to follow scientists and engineers through society.* Milton Keynes: Open University Press.

——— (1988). *The pasteurization of France.* Cambridge, MA: Harvard University Press.

——— (1993). *We have never been modern.* New York: Harvester Wheatsheaf.

——— (1999). On recalling ANT. In J. Law & J. Hassard (Eds.), *Actor network theory and after* (pp. 15–25). Boston, MA: Blackwell Publishers.

Law, J. (2009). Actor network theory and material semiotics. In B. S. Turner (Ed.), *The new Blackwell companion to social theory* (pp. 140–158). Malden, MA: Wiley-Blackwell.

Layton, E. T. J. (1974). Technology as knowledge. *Technology and Culture, 15*(1), 31–41.

Lazzarato. (1996). Immaterial labour. In P. Virno & M. Hardt (Eds.), *Radical thought in Italy: A potential politics* (pp. 133–147). Minneapolis, MN: University of Minnesota Press.

Lenhart, A., Purcell, K., Smith, A., & Zickuhr, K. (2010). Social media and young adults. *The Pew Internet and American Life Project.* Retrieved from The Pew Internet and American Life Project website: www.pewinternet.org/Reports/2010/Social-Media-and-Young-Adults.aspx

Lewchuk, W. 2005, Mass Production. *The Oxford Encyclopedia of Economic History,* ed. J Mokyr, Oxford University Press.

Lorimer, R., Gasher, M., & Skinner, D. (2010). *Mass communication in Canada* (6th ed.). Don Mills, ON: Oxford University Press.

Lyon, D., & Zureik, E. (1996). Surveillance, privacy, and the new technology. In D. Lyon & E. Zureik (Eds.), *Computers, surveillance, and privacy* (pp. 1–18). Minneapolis, MN: University of Minnesota Press.

MacKinnon, R. (2011). Internet wasn't real hero of Egypt. *New America Foundation.* Retrieved from http://articles.cnn.com/2011-02-12/opinion/mackinnon.internet.egypt_1_heroes-printing-press-digital-technologies?_s=PM:OPINION

Mann, S., & Niedzviecki, H. (2001). *Cyborg: Digital destiny and human possibility in the age of the wearable computer.* Toronto: Doubleday Canada.

———, Nolan, J., & Wellman, B. (2003). Surveillance: Inventing and using wearable computing devices for data collection in surveillance environments. *Surveillance & Society, 1*(3), 331–355.

Marcuse, H. (1982). Some social implications of modern technology. In A. Arato & E. Gebhardt (Eds.), *The essential Frankfurt school reader* (pp. 138–162). New York: Continuum.

Markus, L. M., & Robey, D. (1988). Information technology and organizational change: Causal structure in theory and research. *Management Science, 34*(5), 583–598.

Marvin, C. (1988). *When old technologies were new: Thinking about electric communications in the late nineteenth century.* New York: Oxford University Press.

Marx, G. T. (1996). Electric eye in the sky: Some reflections on the new surveillance and popular culture. In D. Lyon & E. Zureik (Eds.), *Computers, surveillance, and privacy* (pp. 193–233). Minneapolis: University of Minnesota Press.

——— (2007). What's new about the "new surveillance"? Classifying for change and continuity. In S. P. Hier & J. Greenberg (Eds.), *The surveillance studies reader* (pp. 83–94). Maidenhead: Open University Press.

Marx, K. (1996). *Das Kapital* (Vol. I). Washington, DC: Gateway Editions.

———, & Engels, F. (1970). *Marx/Engels selected works, volume 3.* Retrieved from www.marxists.org/archive/marx/works/1880/soc-utop/

McCarthy, J. (2007). What is artificial intelligence? *Stanford University.* Retrieved from www.formal.stanford.edu/jmc/whatisai/whatisai.html

McGinn, D. (2011, January 27). Does social media speed up romances? *The Globe and Mail.* Retrieved from www.theglobeandmail.com/life/the-hot-button/does-social-media-speed-up-romances/article1885457/

McGinn, R. E. (1978). What is technology? In P. T. Durbin (Ed.), *Research in philosophy and technology* (Vol. 1, pp. 179–197). Greenwich, CT: Jai Press.

McLuhan, M., McLuhan, E., & Zingrone, F. (1995). *Essential McLuhan.* Don Mills, ON: House of Anansi.

McLuhan, M. (1962). *The Gutenberg galaxy: The making of typographic man.* Toronto, ON: University of Toronto Press.

——— (1964). *Understanding media: The extension of man.* New York: McGraw-Hill.

——— & Powers, B. (1989). *The global village: Transformations in world life and media in the 21st century.* Oxford: Oxford University Press.

———, & Quentin, F. (2003). *The medium is the massage: An inventory of effects.* Toronto, ON: Penguin Canada.

McMullin, J. A. (Ed.). (2011). *Age, gender, and work: Small information technology firms in the new economy.* Vancouver: UBC Press.

McPherson, M., Smith-Lovin, L., & Brashears, M. E. (2006). Social isolation in America: Changes in core discussion networks over two decades. *American Sociological Review, 71*(3), 353–375.

Mehdizadeh, S. (2010). Self-presentation 2.0: Narcissism and self esteem on Facebook. *CyberPsychology, Behavior, and Social Networking, 13*(4), 357–364.

Merton, R. K. (1964). Foreword. In J. Ellul (Ed.), *The technological society* (1st American ed., pp. v–viii). New York: Knopf.

Miller, C. C. (2011, July 19). Google looks for the next Google. Retrieved from www.nytimes.com/2011/07/20/technology/google-spending-millions-to-find-the-next-google.html?pagewanted=all

Mok, D., Wellman, B., & Carrasco, J. A. (2010). Does distance matter in the age of the Internet? *Urban Studies, 47*(13), 2747–2783.

Monge, P. R., & Contractor, N. S. (1997). Emergence of communication networks. In F. M. Jablin & L. L. Putnam (Eds.), *Handbook of organizational communication* (2nd ed.). Thousand Oaks, CA: Sage Publications.

——— & ——— (2003). *Theories of communication networks.* Oxford: Oxford University Press.

Moretti, S. (2010, November 2). Facebook using Canada as testing ground. *London Free Press.* Retrieved from www.lfpress.com/money/2010/11/02/15923101.html

Mossberger, K., Tolbert, C. J., & Stansbury, M. (2003). *Virtual inequality: Beyond the digital divide.* Washington, DC: Georgetown University Press.

Mumford, L. (1967). *The myth of the machine: Technics and human development.* London: Secker & Warburg.

——— (1981). *The culture of cities.* Westport, CT: Greenwood Press.

Naim, M., & Myers, J. J. (2005). Illicit: How smugglers, traffickers and copyrights are hijacking the global economy. *Carnegie Council.* Retrieved from www.carnegiecouncil.org/resources/transcripts/5279.html

National Telecommunications and Information Administration. (2011). Internet and computer use studies and data files. Retrieved from www.ntia.doc.gov/data/index.html

Nielsen. (2009, May 20). Americans watching more TV than ever; web and mobile video up too. Retrieved from http://blog.nielsen.com/nielsenwire/online_mobile/americans-watching-more-tv-than-ever/

Norman, D. A. (1988). *The psychology of everyday things.* New York: Basic Books.

Norris, P. (2001). *Digital divide: Civic engagement, information poverty, and the Internet worldwide.* New York: Cambridge University Press.

Oberg, A., & Walgenbach, P. (2008). Hierarchical structures of communication in a network organization. *Scandinavian Journal of Management, 24,* 183–198.

OECD. (2009). *OECD factbook 2009: Economic, environmental and social statistics.* Retrieved from www.oecd-ilibrary.org/economics/oecd-factbook-2009_factbook-2009-en

Oldenburg, R. (1999). *The great good place: Cafes, coffee shops, community centers, beauty parlors, general stores, bars, hangouts, and how they get you through the day.* New York: Marlowe & Co.

Ollman, B. (1976). *Alienation: Marx's conception of man in capitalist society* (2nd ed.). Cambridge, UK: Cambridge University Press.

One Laptop per Child. (2010a). Education. Retrieved from http://laptop.org/en/vision/education/index.shtml

——— (2010b). Vision. Retrieved from http://laptop.org/en/vision/index.shtml

Ong, W. J. (1991). *Orality and literacy: The technologizing of the word.* London: Routledge.

O'Reilly, T. (2005). What is web 2.0: Design patterns and business models for the next generation of software. Retrieved from http://oreilly.com/pub/a/web2/archive/what-is-web-20.html?page=1

Pacey, A. (1983). *The culture of technology* (1st MIT Press ed.). Cambridge, MA: MIT Press.

Palfrey, J. G., & Gasser, U. (2008). *Born digital: Understanding the first generation of digital natives.* New York: Basic Books.

Pertierra, R., Ugarte, E. F., Pingol, A., Hernandez, J., & Dacanay, N. L. (2003). *Txt-ing selves: Cellphones and Philippine modernity.* Manila: De La Salle University Press.

Pinch, T. (2009). The social construction of technology (SCOT): The old, the new, and the nonhuman. In P. Vannini (Ed.), *Material culture and technology in everyday life: Ethnographic approaches* (pp. 45–58). New York: Peter Lang.

———, & Bijker, W. E. (1987). The social construction of facts and artifacts: Or how the sociology of science and the sociology of technology might benefit each other. *Social Studies of Science, 14*(3), 399–441.

Plummer, T. (2004). Flaked stones and old bones: Biological and cultural evolution at the dawn of technology. *American Journal of Physical Anthropology, 125*(S39), 118–164.

Prensky, M. (2001). Digital natives, digital immigrants. *On the Horizon, 9*(5). Retrieved from www.marcprensky.com/writing/Prensky%20-%20Digital%20Natives,%20Digital%20Immigrants%20-%20Part1.pdf

Privacy Commissioner of Canada. (2009). Facebook agrees to address privacy commissioner's concerns. Retrieved from www.priv.gc.ca/media/nr-c/2009/nr-c_090827_e.cfm

Pruijt, H. D. (1997). *Job design and technology: Taylorism vs. anti-Taylorism.* London: Routledge.

Puma, M. E., Chaplin, D., & Pape, A. D. (2000, September 12). *E-rate and the digital divide: A preliminary analysis from the integrated studies of educational technology.* Retrieved from Urban Institute website: www.urban.org/publications/1000000.html

Putnam, R. D. (2000). *Bowling alone: The collapse and revival of American community.* New York: Simon & Schuster.

Quan-Haase, A. (2009). *Information brokering in the high-tech industry: Online social networks at work.* Berlin: f Publishing.

———, & Collins, J. L. (2008). "I'm there, but I might not want to talk to you": University students' social accessibility in instant messaging. *Information, Communication & Society, 11*(4), 526–543.

———, Cothrel, J., & Wellman, B. (2005). Instant messaging for collaboration: A case study of a high-tech firm. *Journal of Computer-Mediated Communication, 10*(4). Retrieved from http://jcmc.indiana.edu/vol10/issue4/quan-haase.html

———, & Wellman, B. (2004). How does the Internet affect social capital? In M. Huysman & V. Wulf (Eds.), *Social capital and information technology* (pp. 151–176). Cambridge, MA: MIT Press.

——— & ——— (2006). Hyperconnected network: Computer-mediated community in a high-tech organization. In C. Heckscher & P. Adler (Eds.), *The firm as a collaborative community: Reconstructing trust in the knowledge economy* (pp. 281–333). London: Oxford University Press.

——— & ——— (2008). From the computerization movement to computerization: Communication networks in a high-tech organization. In K. L. Kraemer & M. S. Elliott (Eds.), *Computerization movements and technology diffusion: From mainframes to ubiquitous computing* (pp. 203–224). Medford, NJ: Information Today, Inc.

———, & Young, A. L. (2010). Uses and gratifications of social media: A comparison of Facebook and instant messaging. *Bulletin of Science, Technology and Society, 30*(5), 350–361.

Rainie, L., & Wellman, B. (2012). *Networked: The new social operating system.* Cambridge, MA: MIT Press.

Randall, A. J. (1991). *Before the Luddites: Custom, community and machinery in the English woolen industry 1776–1809.* Cambridge: Cambridge University Press.

Rao, A., & Scaruffi, P. (2010). *A history of Silicon Valley: The greatest creation of wealth in the history of the planet.* Palo Alto, CA: Omniware Press.

Ratnagar, S. (2001). The bronze age: Unique instance of a pre-industrial world system? *Current Anthropology, 42*(3), 351–379.

Raynes-Goldie, K. (2010). Aliases, creeping, and wall cleaning: Understanding privacy in the age of Facebook. *First Monday, 15*(1), January. Retrieved from http://firstmonday.org/htbin/cgiwrap/bin/ojs/index.php/fm/article/viewArticle/2775

Reese, L., Faist, J., & Sands, G. (2010). Measuring the creative class. *Journal of Urban Affairs, 32*(3), 345–366.

Regional Surveys of the World. (2004). *The Far East and Australia* (35 ed.). London: Europa Publications, Taylor and Francis Group.

Rentschler, E. (1996). *The ministry of illusion: Nazi cinema and its afterlife*. Cambridge, MA: Harvard University Press.

Restivo, S. P. (2005). *Science, technology, and society: An encyclopedia*. Oxford, UK: Oxford University Press.

Rheingold, H. (2000). *The virtual community: Homesteading on the electronic frontier* (Rev. ed.). Cambridge, MA: MIT Press.

Rich, L. (2010). Shiny new things. *Ad Age Insights*. Retrieved from http://adage.com/images/bin/pdf/shiny_new_things.pdf

Roberts, J. (2007, July 6). AK-47 inventor says conscience is clear. *CBS News*. Retrieved from www.cbsnews.com/stories/2007/07/06/world/main3025193.shtml?source=RSSattr=World_3025193%29

Roebuck, J. (1982). *The making of modern English society from 1850* (2nd ed.). London: Routledge.

Roehrs, J. (1998). *A study of social organization in science in the age of computer-mediated communication*. Unpublished doctoral thesis, Nova Southeastern University.

Rogers, E. M. (2003). *Diffusion of innovations* (5th ed.). New York: Free Press.

Rogers, E. M. (1983). *Diffusion of innovations* (3rd ed.). New York, NY: Free Press.

Rogers, T. W. (1994). Detournement for fun and (political) profit. *Ctheory*. Retrieved from http://ctheory.net/text_file.asp?pick=242

Rohrbeck, R., Döhler, M., & Arnold, H. M. (2009). Creating growth with externalization of R&D results-the spin-along approach. *Global Business and Organizational Excellence, 28*(4), 44–51. Retrieved from http://papers.ssrn.com/sol3/papers.cfm?abstract_id=1472133

Rubin, V. L., Chen, Y., & Thorimbert, L. M. (2010). Artificially intelligent conversational agents in libraries. *Library Hi Tech, 28*(4), 496–522.

Russell, R. (2011, July 26). An inconvenient truth about Toyota. *The Globe and Mail*. Retrieved from http://m.theglobeandmail.com/globe-drive/car-tips/safety/an-inconvenient-truth-about-toyota/article2104879/?service=mobile

Russell, S. J., & Norvig, P. (2003). *Artificial intelligence: A modern approach* (2nd ed.). Upper Saddle River, NJ: Prentice Hall.

Sacks, D. (2011, January 31). How YouTube's global platform is redefining the entertainment business. *Fast Company*. Retrieved from www.fastcompany.com/magazine/152/blown-away.html

Sale, K. (1995). *Rebels against the future: The Luddites and their war on the industrial revolution: Lessons for the computer age*. Reading, MA: Addison-Wesley.

Sanderson, J., & Cheong, P. H. (2010). Tweeting prayers and communicating grief over Michael Jackson online. *Bulletin of Science, Technology and Society, 30*(5), 328–340. Retrieved from http://bst.sagepub.com/content/30/5/328.full.pdf+html

Schummer, J. (2006). Societal and ethical implications of nanotechnology: Meanings, interest groups and social dynamics. In J. Schummer & D. Baird (Eds.), *Nanotechnology challenges: Implications for philosophy, ethics and society* (pp. 413–449). London: World Scientific Publishing.

Schumpeter, J. A. (2004). *The theory of economic development: An inquiry into profits, capital, credit, interest, and the business cycle* (1934 ed.). New Brunswick, NJ: Transaction Publishers.

Sciadas, G. (2002). *The digital divide in Canada*. Science Innovation and Electronic Information Division. Retrieved from http://dsp-psd.pwgsc.gc.ca/Collection/Statcan/56F0009X/56F0009XIE2002001.pdf

Sears, A., & Jacko, J. A. (2008). *The human-computer interaction handbook: Fundamentals, evolving technologies, and emerging applications* (2nd ed.). New York: Lawrence Erlbaum Associates.

Segal, H. P. (1985). *Technological utopianism in American culture*. Chicago: University of Chicago Press.

Sellen, A., & Harper, R. H. R. (2002). *The myth of the paperless office*. Cambridge: MIT Press.

Servaes, J. (1999). *Communication for development: One world, multiple cultures*. Cresskill, NJ: Hampton.

Servaes, J. (2010). Social change. *Oxford Bibliographies Online*. Retrieved from www.oxfordbibliographiesonline.com/view/document/obo-9780199756841/obo-9780199756841-0063.xml;jsessionid=8D56E1250611451078B7224F60235FC6

Simpson, L. C. (1995). *Technology, time, and the conversations of modernity*. New York: Routledge.

Smith, A., Campbell, R. H., & Skinner, A. S. (1981). *An inquiry into the nature and causes of the wealth of nations*. Indianapolis: Liberty Classics.

Smith, S. (2005, December 1). The $100 laptop—is it a wind-up? Retrieved from http://edition.cnn.com/2005/WORLD/africa/12/01/laptop/

Snezhurov, Y. (2011, April 12). Museum Kalashnikov. Retrieved from www.kalashnikov-museum.udmnet.ru/kalash5e.htm

Social Bakers. (2011, March 17). Interesting digital marketing trends in the Middle East. Retrieved from www.socialbakers.com/blog/130-interesting-digital-marketing-trends-in-the-middle-east/

Sofka, C. J. (1997). Social support "Internetworks," Caskets for sale, and more: Thanatology and the information superhighway. *Death Studies, 21*(6), 553–574.

Sorensen, C. (2010, July 29). Toyota's latest repairs. *Maclean's*. Retrieved from www2.macleans.ca/2010/07/29/toyotas-latest-repairs/

Sproull, L. S., & Kiesler, S. B. (1991). *Connections: New ways of working in the networked organization.* Cambridge, MA: MIT Press.

Srinivasan, R. (2011, February 15). The net worth of open networks. *The Huffington Post*. Retrieved from www.huffingtonpost.com/ramesh-srinivasan/the-net-worth-of-open-net_b_823570.html

Statistics Canada. (2009). *Canadian Internet use survey 2009.* CANSIM. Retrieved from www.statcan.gc.ca/daily-quotidien/100510/dq100510a-eng.htm

Stedmon, A. W. (2011). The camera never lies, or does it? The dangers of taking CCTV surveillance at face-value. *Surveillance & Society, 8*(4), 1477–7487.

Steiner, G.A. (1963). *The people look at television: A study of audience attitudes.* New York: Knopf.

Stevenson, S. (2009). Digital divide: A discursive move away from the real inequities. *The Information Society, 25*(1), 1–22.

Stone, B. (2007). Microsoft buys stake in Facebook. *The New York Times, Technology*. Retrieved from www.nytimes.com/2007/10/25/technology/25facebook.html

Street, J. (1992). *Politics and technology.* New York: Guilford Press.

Sundén, J. (2003). *Material virtualities: Approaching online textual embodiment.* New York: Peter Lang Publishing.

Sundström, P. (1998). Interpreting the notion that technology is value-neutral? *Medicine, Health Care and Philosophy, 1*, 41–45.

Sunstein, C. R. (2001). *Republic.Com.* Princeton, NJ: Princeton University Press.

Sward, K. (1948). *The legend of Henry Ford.* New York: Rinehart.

Swartz, J. (2011, April 21). Tech jobs boom like it's 1999. *USA Today*. Retrieved from www.usatoday.com/tech/news/2011-04-20-tech-jobs-booming.htm

Tapscott, D., & Williams, A. D. (2006). *Wikinomics: How mass collaboration changes everything.* New York: Portfolio.

Tenner, E. (2003). *Our own devices: The past and future of body technology.* New York: Alfred A. Knopf.

Tenner, E. (1996). *Why things bite back: Technology and the revenge of unintended consequences.* New York: Knopf.

The Canadian Press. (2008, December 31). Canadian Internet usage at home in 2007. CBC. Retrieved from www.cbc.ca/news/interactives/cp-internet-use/

The real digital divide. (2005, March 10). *The Economist*. Retrieved from www.economist.com/node/3742817?story_id=3742817

Toffler, A. (1970). *Future shock.* New York: Random House.

——— (1980). *The third wave.* New York: Morrow.

Tönnies, F. (2004). *Community & society. (Gemeinschaft und Gesellschaft).* New Brunswick, NJ: Michigan State University Press.

Trimborn, J. (2007). *Leni Riefenstahl: A life.* New York: Faber and Faber.

Tufekci, Z. (2007). Can you see me now? Audience and disclosure regulation in online social network sites. *Bulletin of Science, Technology & Society, 28*(1), 20–36.

Tufekci, Z. (2008). Grooming, gossip, Facebook and MySpace. *Information, Communication and Society, 11*, 544–564.

——— (2010). Who acquires friends through social media and why? "Rich get richer" versus "seek and ye shall find." Presented at 4th Int'l AAAI Conference on Weblogs and Social Media (ICWSM, 2010). ICWSM '09. (Washington, DC, May 23–26: AAAI Press).

Turing, A. M. (1950). Computing machinery and intelligence. *Mind, 59*, 433–446.

Van Dijck, J. (2007). *Mediated memories in the digital age.* Stanford, CA: Stanford University Press.

Vehovar, V., Sicherl, P., Husing, T., & Donicar, V. (2009). Methodological challenges of digital divide measurements. *The Information Society, 22*, 279–290.

Verbeek, P.-P. (2005). *What things do: Philosophical reflections on technology, agency, and design.* University Park, PA: Pennsylvania State University Press.

Viseu, A. (2002). *Augmented bodies and behavior bias interfaces.* Proceedings for the Society for Social Studies of Science, Milwaukee,

WI. Retrieved from www.yorku.ca/aviseu/ pdf%20files/viseu_4S2002.pdf

————, Clement, A., & Aspinall, J. (2004). Situating privacy online: Complex perceptions and everyday practices. *Information, Communication & Society, 7*(1), pp. 92–114.

Walsham, G. (1997). Actor-network theory and its research: Current status and future prospects. In J. I. DeGross, A. S. Lee & J. Liebenau (Eds.), *Information systems and qualitative research: Proceedings of the IFIP TC8 WG 8.2 international conference on information systems and qualitative research, 31st May–3rd June 1997, Philadelphia, Pennsylvania, USA* (1st ed., pp. 466–480). London: Chapman & Hall.

Wang, H., & Wellman, B. (2010). Social connectivity in America: Changes in adult friendship network size from 2002 to 2007. *American Behavioral Scientist, 53*(8), 1148–1169.

Weber, M. (1920/2003). *The Protestant ethic and the spirit of capitalism* (Rev. 1920 ed.). Mineola, New York: Dover Publications.

Weinstein, J. (2010). *Social change* (3rd ed.). Lanham, MD: Rowman & Littlefield.

Wellman, B. (1979). The community question. *American Journal of Sociology, 84*, 1201–1231.

————, & Berkowitz, S. D. (Eds.). (1988). *Social structures: A network approach.* Cambridge: Cambridge University Press.

———— (2001). Physical place and cyber place: The rise of personalized networking. *International Journal of Urban and Regional Research, 25*(2), 227–252.

————, & Frank, K. (2001). Network capital in a multilevel world: Getting support from personal communities. In N. Lin, K. Cook & R. S. Burt (Eds.), *Social capital: Theory and research* (pp. 233–273). Hawthorne, NY: Aldine de Gruyter.

————, & Haythornthwaite, C. (2002). Introduction. In B. Wellman & C. Haythornthwaite (Eds.), *The Internet in everyday life.* Oxford: Blackwell Publishers.

————, Quan-Haase, A., Witte, J., & Hampton, K. (2001). Does the Internet increase, decrease, or supplement social capital? Social networks, participation, and community commitment. *American Behavioral Scientist, 45*(3), 437–456.

————, & Wortley, S. (1990). Different strokes from different folks: Community ties and social support. *American Journal of Sociology, 96*(3), 558–588.

Westin, A. F. (2003). Social and political dimensions of privacy. *Journal of Social Issues, 59*(2), 431–453

Wikipedia. (2011). Neutral point of view. Retrieved from http://en.wikipedia.org/wiki/Wikipedia:Neutral_point_of_view

Winner, L. (1999). Do artifacts have politics? In D. Mackenzie & J. Wajcman (Eds.), *The social shaping of technology* (pp. 28–40). Buckingham: Open University Press.

———— (2003). Social constructivism: Opening the black box and finding it empty. In R. C. Scharff & V. Dusek (Eds.), *Philosophy of technology: The technological condition: An anthology* (pp. 233–244). Malden, MA: Blackwell Publishers.

Wysocki, R. K. (2006). *Effective software project management.* Indianapolis, IN: Wiley Publishing.

Young, A. L. (2008). *Defacing the 'book: Examining information revelation, Internet privacy concerns and privacy protection in Facebook* (Unpublished master's thesis). University of Western Ontario, London, ON.

———— & Quan-Haase, A. (2009). Information revelation and Internet privacy concerns on social network sites: A case study of Facebook. In J. M. Carrol (Ed.). *Fourth International Conference on Communities and Technologies (University Park, PA, USA, June 25–27)*, pp. 265–274. Dordrecht: Springer Verlag.

Zachary, P. G. (1998). Armed truce: Software in an age of teams. *Information, Technology & People, 11*(1), 62–65.

Zimmerman, M. E. (1990). *Heidegger's confrontation with modernity: Technology, politics and art.* Bloomington, IN: Indiana University Press.

Zuckerman, E. (2011, January 14). The first Twitter revolution? *Foreign Policy.* Retrieved from www.foreignpolicy.com/articles/2011/01/14/the_first_twitter_revolution?page=full

Index

Note: Page numbers in italics indicate figures.